SpringerBriefs in Optimization

Series Editors

Panos M. Pardalos
János D. Pintér
Stephen M. Robinson
Tamás Terlaky
My T. Thai

SpringerBriefs in Optimization showcases algorithmic and theoretical techniques, case studies, and applications within the broad-based field of optimization. Manuscripts related to the ever-growing applications of optimization in applied mathematics, engineering, medicine, economics, and other applied sciences are encouraged.

For further volumes:
http://www.springer.com/series/8918

Leping Yang • Yanwei Zhu • Xianhai Ren
Yuanwen Zhang

On-Orbit Operations Optimization

Modeling and Algorithms

 Springer

Leping Yang
College of Aerospace Science
  and Engineering
National University of Defense
  Technology
Changsha, Hunan, China

Yanwei Zhu
College of Aerospace Science
  and Engineering
National University of Defense
  Technology
Changsha, Hunan, China

Xianhai Ren
PLA Airforce
Wuxi, Jiangsu, China

Yuanwen Zhang
College of Aerospace Science
  and Engineering
National University of Defense
  Technology
Changsha, Hunan, China

ISSN 2190-8354   ISSN 2191-575X (electronic)
ISBN 978-1-4939-0837-0 ISBN 978-1-4939-0838-7 (eBook)
DOI 10.1007/978-1-4939-0838-7
Springer New York Heidelberg Dordrecht London

Library of Congress Control Number: 2014936414

Mathematics Subject Classification: 37E15, 46A55, 49-02, 49J30, 49L20, 70E55

Printed on acid-free paper

Springer is part of Springer Science+Business Media (www.springer.com)

Preface

On-orbit operations generally refer to the spacecraft undergoing or executing numerous events, including parameter configuration, system maintenance, attitude orienting, and orbital maneuver. On-orbit operations optimization aims to provide optimized space services or products, which may be more cost-effective, or more responsive, or having better performance. Recently, with on-orbit servicing development, interest in the operations optimization among multiple cooperative or noncooperative spacecraft has grown. As a result, on-orbit servicing operations mission planning (optimization) emerges as a hot topic. Despite numerous journal and conference articles in the topic, so far there has been no published monograph dedicated to the subject. Based on two new doctoral dissertations of our research group, this book summarizes latest optimization models and algorithms of on-orbit servicing operations, to present a ready reference for both researchers and engineers.

On-orbit servicing operations involve three key technology areas: rendezvous and docking, manipulation, and autonomy, which are all closely related to optimization theory and methodology. Rendezvous and docking mission design and analysis are essentially an optimization problem of spacecraft trajectory. Manipulator design and control can be seen as an optimization problem based on complex multi-body dynamics and coordinated control algorithms. And autonomous operations must be performed on the basis of intelligent optimization algorithms, which is a specific application of on-orbit servicing mission optimization. To sum up, optimization problems are common in the development and application of on-orbit servicing technology, and optimization theory and methodology provide fundamental tools for on-orbit servicing mission design and analysis.

On-orbit servicing operations optimization is an extension of classic optimization problem of spacecraft trajectory. While the basic tools of optimization theory such as the calculus of variations, Pontryagin's principle, and Hamilton-Jacobi theory have not changed, there has been a revolution in the manner in which they are applied and in the development of numerical optimization due to more complex dynamics and constraints. Summarizing our recent research work, this book investigates spacecraft multi-mission planning, far-range orbital maneuver planning, proximity relative motion planning, and multi-spacecraft coordinated planning, building up a modeling and algorithm framework for on-orbit servicing operations optimization.

It should be honestly pointed out that this book is far from complete and comprehensive. The continuing emergence of new optimization problems requires consistent investigation and novel algorithms. Generally, as on-orbit operations optimization problems grow in complexity, numerical solution method seems to be more preferred than analytical method, and one of the future focuses may be the direct method that transforms the continuous optimal control problem to the parameter optimization one.

Changsha, Hunan, China   Leping Yang
Changsha, Hunan, China   Yanwei Zhu
Wuxi, Jiangsu, China   Xianhai Ren
Changsha, Hunan, China   Yuanwen Zhang

Acknowledgments

This book is a joint research product mainly based on two new doctoral dissertations of our research group, and also accredited to the researchers of our reference source. The book is attributable to these collaborators, and is an attempt to distill these contributions into a manageable and digestible form. We are forever indebted to our colleagues and students, without whom this book and the research on which it is based could not exist.

Firstly, we want to express our gratitude to Mr. Zhang Weihua and Mr. Lv Quanqing, senior officials of our college, who give great support and care to our research team. We would like to acknowledge Ms. Peng Wangqiong, who helps with the administrative routines and as an English major makes outstanding contributions in the preparation of the manuscript. Special thanks should go to Mr. Zhang Zhenmin, who gives great support to our research programs. We would also like to pay our gratitude to Mr. Zhang Qingbin and Mr. Liu Xinjian, associate professors who provide many enlightening ideas in our academic discussions. We are also grateful to several insightful and energetic students, including Huang Huan, Cai Weiwei, and Liu Haitao, who help with the simulation cases and manuscript formatting. Finally, the work in this book is partially supported by National High-Tech Research and Development Program and National Natural Science Foundation of China (grant. 11172322).

Contents

List of Abbreviations

DSS   Distributed Satellite System
ECI   Earth Centered Inertia
FOV   Field of View
GA   Genetic Algorithm
HGABB   Hybrid Genetic Algorithm Branch and Bound
hp-APM   *hp*-Adaptive Pseudospectral Method
IAPF   Improved Artificial Potential Function
IDVD   Inverse Dynamics in the Virtual Domain method
LOS   Line of Sight
LP   Linear Programming
MILP   Mixed Integer Linear Programming
MOEA   Multi-Objective Evolutionary Algorithm
MOEA/D   Multi-Objective Evolutionary Algorithm based on Decomposition
MOOP   Multi-Objective Optimization Problem
MRP   Modified Rodrigues Parameters
NLP   Nonlinear Programming
NPGA   Niched Pareto Genetic Algorithm
OOS   On-Orbit Servicing
RPM   Radau Pseudospectral Method
RVD   Rendezvous and Docking
SOOP   Single Objective Optimization Problem
SQP   Sequential Quadratic Programming
TT&C   Tracking, Telemetering, and Command
VEGA   Vector Evaluated Genetic Algorithm

Chapter 1
Introduction

1.1 Background

Space missions serve the purpose to deliver mission products in response to users' requests. As an important part of space missions, on-orbit operations are decisive to space mission success and efficiency. In the past two decades, with the successful accomplishment of a series of Hubble Space Telescope repairs, assembly of the International Space Station, as well as demonstration of the Orbital Express, the growing importance of On-Orbit Servicing (OOS) has been emphasized. OOS, which has grown to be a hot topic in on-orbit operations field, will help transit space utilization into a new era of operational efficiency and cost-effectiveness.

1.1.1 On-Orbit Servicing Concept

OOS is the process of improving a space-based capability through a combination of orbital activities that may include inspection, rendezvous and docking (RVD), and value-added modifications to a spacecraft's position, orientation, and operational status. The OOS activities can be categorized into five high-level functions [1–3]:

1. Inspect. Observation of the customer spacecraft (customer) from an attached position to assess its physical and operational status, and may be a necessary precursor for other OOS activities.
2. Relocate. Modification of the customer orbit to support constellation reconfiguration, tactical maneuver, and de-orbit or rescue.
3. Restore. Returning the customer to a previous state or intended state to enable a wide range of capabilities. Typical restoration activities include refueling, maneuvering, and repairing faulty hardware for lifetime extension.

L. Yang et al., *On-Orbit Operations Optimization: Modeling and Algorithms*,
SpringerBriefs in Optimization, DOI 10.1007/978-1-4939-0838-7_1,
© Leping Yang, Yanwei Zhu, Xianhai Ren, Yuanwen Zhang 2014

4. Augment. Increasing the capability of the customer by replacing or adding hardware to improve spacecraft performance.
5. Assemble. Mating modules to enable the construction of large space platforms that may not be transported by existing launch vehicles.

OOS benefits can be summarized as reducing risk of mission failure and cost, enhancing mission performance and flexibility, and enabling new missions.

1.1.2 Key Technology Areas

From the view of the servicer spacecraft (servicer), OOS operations involve three key technology areas: RVD, manipulation, and autonomy.

RVD is separated into three sequential phases: rendezvous, proximity operations, and docking. RVD with cooperative, non-spinning customers is already a mature technology. Docking becomes more complex if the customer is spinning or tumbling out of control. Control algorithms must determine the spin rate, spin axis orientation, and any nutation. Proximity operations must be planned and executed to maneuver the servicer along the customer's spin axis and match spin rates.

Manipulators, also known as "robot arms", are necessary for the physical interaction between spacecraft. Basic teleoperation of manipulators is well established; however, finer, more precise motions and operation in more constrained work zones for OOS are less mature. Space robotics differs from terrestrial applications in the structural characteristics of manipulators and payloads in zero gravity. Therefore, control algorithms for manipulators must take into account flexible multi-body dynamics and contact dynamics.

Robot control sophistication ranges from teleoperation, to automatic operation, and to increasing levels of autonomy. Autonomy is required only when the task is unfamiliar or complex, or when communication issues exist, whose benefits include improvements in efficiency, robustness, and capability. At higher levels of autonomy, artificial intelligence algorithms enable the robot to learn as it operates and improve task planning for future activities.

In conclusion, optimization problems are common in RVD, manipulation, and autonomy, highlighting the need to study optimization theory and algorithms.

1.2 On-Orbit Servicing Operations

In a typical OOS mission, the servicer completes a correlated mission sequence from being launched to the orbit, performing predetermined service, to demating from the customer and preparing for the next service. Seven specific operation steps are listed below [3].

1. Transfer to the orbit and checkout. Transfer and checkout are the first step to complete OOS operations, involving the servicer to achieve the necessary orbital plane and proper position to interact with the customer, and performing predetermined checkout for all systems to prepare rendezvous.
2. Rendezvous with the customer. Rendezvous is the series of activities occurring at relative positions approximately between 300 km and 1 km apart, whose goal is to approach the customer as quickly, efficiently, and safely as possible in order to prepare for proximity operations.
3. Proximity and approach operations. Proximity and approach operations are the series of activities occurring at relative positions approximately between 1 km and 30 m apart, whose goal is to close the relative position between the two spacecraft as safely as possible, given the increased risk of collision while performing final checks to prepare for grappling.
4. Grapple with customer. Grappling and subsequent activities that are performed while grappled are the key operations in the OOS mission. The term "grappling" refers to the mating or joining of the two spacecraft. Activities performed while the two spacecraft are grappled include attitude control, communications relay, and inspection. These functions, combined with the servicing functions, are referred to as mated operations.
5. Perform servicing operations. Servicing operations are the activities that occur between the mated operations and the eventual separation and/or disposal of the spacecraft, which represents the purpose of the servicing mission. The typical servicing operations include relocation, mechanical assist, repair/upgrade, resource replenishment, and assembly for the customer.
6. Demate from customer. The demating operation is essentially the inverse of the grappling operation in that the connecting interfaces between the servicer and customer are terminated. This may occur in two phases—the relaxation of the rigidized connection followed by the release of the temporary connection.
7. Transfer to next service. One possible option following the demating phase is for the servicer to make preparations for the next servicing mission, which results in returning to the starting state of the mission sequence with the servicer having translated to a pre-rendezvous orbital position for the new customer. Another possible option is to make preparations to loiter in a temporary parking orbit while waiting for additional servicing opportunities.

1.3   Optimization Problem

The complexity of OOS operations optimization is reflected by multidimensional, correlated and embedded constraints, nonlinear, flexible multi-body dynamics, and feasible and appropriate optimization algorithms. Given assured security, most common optimization criteria are least propellant consumption, shortest maneuver time, or best maneuver path. Constraints, dynamics, and algorithms are the key elements for on-orbit operations optimization problem.

1.3.1 Constraints

For a typical on-orbit operations mission, several kinds of constraints need to be considered, categorized as follows.

1. Sensor
   CCD camera is the most common sensor for measurement and navigation, whose operation constraints are usually formulated as line of sight (LOS), field of view (FOV), and lighting constraints, posing limitations on inspection angle and distance.
2. Controller
   Controller constraints, which are usually characterized by the mode, magnitude, and configuration of the thruster, determine the approach to formulate and solve the on-orbit operations optimization problem.
3. State
   State constraints, i.e., the limitations on relative position and attitude, include collision avoidance, plume contamination, mission limitation, and passive safety, among which mission limitation refers to no-fly zones or envelopes determined by FOV based on customer characteristics, and passive safety is essentially collision avoidance in the event of thruster failures, computer anomalies, and loss of sensing.

In addition, considering the requirements to maintain power positive mode and spacecraft components within operational and/or survival temperature limits, Earth shadow constraint must be addressed.

Note that not all constraints above are considered in a specific mission and the selection of constraints determines model complexity and feasible algorithm.

1.3.2 Dynamics

OOS operations require that the servicer rendezvous with the customer, perform proximity operations, then capture or berth to, or dock with the customer. These activities are part of the discipline of astrodynamics, in which the spacecraft performs an intricate "dance" under the influences of gravity and thrusters [3].

In OOS optimization, three kinds of dynamic models, i.e., absolute motion dynamic model for far-range rendezvous, relative motion dynamic model for proximity and approach, and multi-body dynamic model considering rigid-flexible coupling effects for the manipulator-enabled grappling and servicing operations, are usually exploited. Among these models, the first one builds the base for the last two and can be formulated as the Newton's equations or the classical orbital elements, where the former has the continuous mathematical expression while the latter is physically meaningful, and both are applied in Chaps. 2 and 3 of this book. As for the relative dynamic model, which can be formulated by dynamics method or kinematics method, there are many equation formats derived from different methods, hypotheses of gravitational force and reference orbital motion. In Chaps. 4 and 5, several relative motion dynamic models are applied, including T-H, C-W equations, etc., selected by feasibility yielded by the relative distance between spacecraft and the reference orbital motion type. The multi-body dynamic model could be formulated by such methods as the Newton's mechanics method, the Lagrange's equations, the Kane's method, etc., which are primarily concerned with manipulator flexibility and its dynamic coupling with the base. However, we do not elaborate this due to the scope limit of this book.

1.3.3 Algorithms

The optimization algorithms can be categorized as analytical approaches and numerical solution algorithms. The analytical approaches are based on the necessary conditions for optimality that are derived through calculus of variations. The numerical solution algorithms are further categorized as indirect and direct. The indirect methods transcribe the original optimization control problem into a Hamilton boundary value problem by utilizing the well-known Pontryagin's Maximum Principle and introducing co-state variables to express the optimal control variables. The direct methods transcribe a continuous optimization control problem into a parameter optimization one via discretization, and then solve the resulting nonlinear programming (NLP) problem.

The analytical approaches are only suitable for certain simplified or special cases due to their inherent limitations. The indirect methods provide high accuracy solutions, yet the deducing process is complicated, the performance is strongly dependent on a good initial guess for states and co-states, and it is difficult to solve the problem with path constraints. The direct methods have the advantages of better convergence, insensitiveness to initial guesses, and freedom of deducing the necessary conditions, yet the resulting NLP may cause heavy computation load and difficulty in guaranteeing the optimality of the obtained solutions. One principal way by which direct methods are distinguished is with regard to what variables are discretized, only control variables, only state variables, or both [4–6]. Among various direct methods, the collocation method and pseudospectral method are perhaps the best known and most implemented methods for on-orbit operations optimization.

In addition, some intelligent algorithms like evolutionary algorithms, simulated annealing, and particle swarm optimization have been extensively applied recently. The most prominent advantage of intelligent algorithms is that they are more likely to locate a global minimum in the search space rather than be attached to a local minimum.

1.4   Outline of the Book

The remainder of this book consists of four chapters.

Chapter 2 deals with spacecraft multi-mission planning. We establish a multilevel nonlinear planning model, which is then simplified as mission assignment, sequence arrangement, and time distribution. Then, for N-to-N spacecraft mission assignment, the integer programming algorithm is adopted, while for one-to-N sequence arrangement and time distribution, the Hybrid Genetic Algorithm Branch and Bound (HGABB) algorithm is explored.

Chapter 3 investigates far-range orbital maneuver planning which aims to optimize the transfer trajectory for the servicer under the constraints of time, fuel, and observation and control. Given the impulse thruster mode, on the basis of Lambert solution, we adopt the Genetic Algorithm (GA) and randomized A* tree expansion algorithm to generate the optimized multi-impulse transfer trajectory respectively.

Chapter 4 studies proximity relative motion planning, which is further divided into common and close cases, based on spacecraft relative motion dynamics. For the former, we examine the trajectory planning models and algorithms specific to impulse thrust, Bang-Bang thrust, and constant low thrust respectively. For the latter, in a specific close proximity inspection mission, the *hp*-Adaptive Pseudospectral Method (*hp*-APM) is adopted if the customer is free; the Improved Artificial Potential Function (IAPF) is exploited if the FOV is blocked; and the Inverse Dynamics in the Virtual Domain method (IDVD) is explored if the customer is maneuvered.

Chapter 5 primarily concentrates on multi-spacecraft coordinated planning. The cyclic pursuit method and the contraction theory are investigated respectively. The former is illustrated by cases of constrained circular flyaround, natural elliptical flyaround, flyaround with different orbit, and cubic flyaround, while the latter by a four-spacecraft cluster case.

References

1. International Space University. (2007). *DOCTOR: Developing on-orbit servicing concepts, technology options, and roadmap*. Final report of the International Space University. Summer Session Program.
2. Long, A., Richards, M., & Hastings, D. (2007). On-orbit servicing: A new value proposition for satellite design and operation. *AIAA Journal of Spacecraft and Rockets, 44*(4), 964–975.
3. National Aeronautics and Space Administration Goddard Space Flight Center. (2010). On-orbit satellite servicing study project report.
4. Ben-Asher, J. (2009). *Optimal control theory with aerospace applications*. Reston, VA: AIAA.
5. Gong, Q., Ross, I., Kang, W., & Fahroo, F. (2008). Connections between the convector mapping theorem and convergence of pseudospectral methods for optimal control. *Computational Optimization and Application, 41*, 307–335.
6. Tan, K. C., Khor, E. F., & Lee, T. H. (2005). *Multiobjective evolutionary algorithms and applications*. London: Springer.

Chapter 2
Spacecraft Multi-Mission Planning

Abstract For on-orbit operations missions, the multi-mission mode of one or a group of servicers operating on multiple customers will gradually come true. Spacecraft multi-mission planning is founded on one-to-one single mission planning, including far-range orbital maneuver planning and proximity relative motion planning. In this chapter, a multilevel nonlinear planning model, which can be simplified as mission assignment, sequence arrangement, and time distribution, is established and solution strategy explored. For N-to-N spacecraft mission assignment, we assume one servicer only serves one customer, and the integer programming algorithm is adopted. For one-to-N sequence arrangement and time distribution, the HGABB combined by GA and the branch and bound algorithm is applied to a multiple GEO satellites flyaround inspection scenario.

2.1 Problem Formulation

Assume the customers gather in a constellation or in a certain orbit, and the servicers are deployed as a constellation. Since the customers differ in criticality and service imminence, their mission priority also varies. One servicer can serve one or several customers, constrained by mission duration, fuel consumption, and observation and control.

Spacecraft multi-mission planning helps select appropriate servicer for customer so that the missions with higher priority are first addressed; meanwhile time and fuel costs are optimized. First, mission assignment is optimized to generate an objective set for each servicer. Second, mission sequence is such planned that every servicer is appropriately assigned. Third, time distribution is optimized so that the total fuel consumption to serve all customers is minimized under a given sequence.

L. Yang et al., *On-Orbit Operations Optimization: Modeling and Algorithms*,
SpringerBriefs in Optimization, DOI 10.1007/978-1-4939-0838-7_2,
© Leping Yang, Yanwei Zhu, Xianhai Ren, Yuanwen Zhang 2014

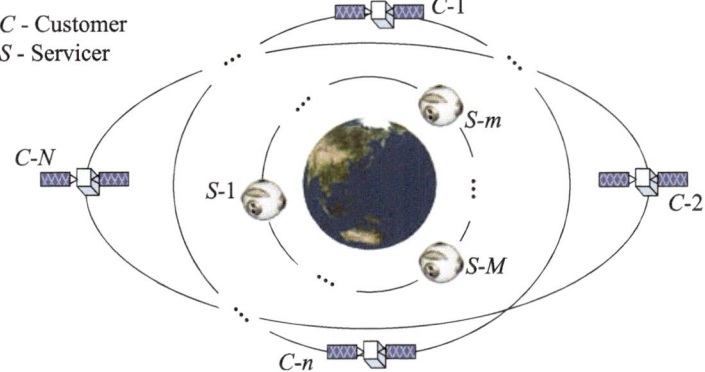

Fig. 2.1 Spacecraft multi-mission planning scenario

To sum up, spacecraft multi-mission planning can be treated as a multi-objective multilevel nonlinear planning problem.

Since time and fuel consumption are much greater in the far-range orbital maneuver phase as compared to the proximity operation phase, the calculation of cost function is based on the far-range orbital maneuver phase here.

2.1.1 Planning Model

In general, the multi-mission planning scenario is defined as in Fig. 2.1, where there are N customers and M servicers ($M \leq N$). A mathematical model is then established as below [1].

2.1.1.1 Decision Variables

For spacecraft multi-mission planning, three kinds of decision variables need to be defined.

1. Mission assignment decision variables
   The mission assignment decision variables are defined as

$$\mathbf{X} = \begin{bmatrix} x_{11} & \cdots & x_{1N} \\ \vdots & \ddots & \vdots \\ x_{M1} & \cdots & x_{MN} \end{bmatrix} \qquad (2.1)$$

where

$$x_{mn} = \begin{cases} 1 & \text{assign servicer } m \text{ to serve customer } n \\ 0 & \text{otherwise} \end{cases} \quad m = 1, \ldots, M; \ n = 1, \ldots, N \tag{2.2}$$

by which mission assignment is transformed to a 0–1 integer programming problem.

2. Mission sequence decision variables

The mission sequence decision variables are defined as

$$\mathbf{Q} = [Q_1 \quad \cdots \quad Q_M]^{\mathrm{T}}, \ Q_m = \{s_{m1}, s_{m2}, \ldots, s_{ma_m}\} \tag{2.3}$$

where Q_m refers to the mission sequence set of servicer m, a_m the number of missions assigned to servicer m, and $s_{mk}(k = 1, \ldots, a_m)$ the index of customer when servicer m performs the kth mission.

3. Time distribution decision variables

The time distribution decision variables are defined as

$$\mathbf{T} = [T_1 \quad \cdots \quad T_M]^{\mathrm{T}}, \ T_m = [\Delta t_{m1} \quad \Delta t_{m2} \quad \cdots \quad \Delta t_{ma_m}] \tag{2.4}$$

where T_m denotes the time distribution scheme of Q_m, and Δt_{mk} the duration of the kth mission.

2.1.1.2 Cost Function

According to the mission requirements, four optimization criteria, maximum priority, minimum fuel, minimum time, and fuel consumption equilibrium among servicers, are selected.

Defining p_n as the mission priority of customer n, the cost function of maximum priority can be written as

$$\max J_1 = \sum_{m=1}^{M} \sum_{n=1}^{N} x_{mn} p_n \tag{2.5}$$

Defining ΔV_{mk} as the velocity increment for servicer m to finish the kth mission, ΔV_m the total velocity increment for servicer m to finish the mission sequence Q_m, the cost function of minimum fuel can be written as

$$\min J_2 = \Delta V = \sum_{m=1}^{M} (\Delta V_m) = \sum_{m=1}^{M} \left(\sum_{k=1}^{a_m} \Delta V_{mk} \right) \tag{2.6}$$

where ΔV_{mk} could be solved by the two-impulse multiple-revolution Lambert transfer algorithm (see Sect. 3.1.1).

Defining Δt_m as the total time for servicer m to finish the mission sequence Q_m, then $\Delta t_m = \sum\limits_{k=1}^{a_m} \Delta t_{mk}$, the cost function of minimum time can be written as

$$\min J_3 = \max\{\Delta t_m\} \tag{2.7}$$

Defining C_m as the remaining orbital maneuver capability, indicated by the velocity increment, and $\overline{C}$ as the mean value of all servicers, then $\overline{C} = \sum\limits_{m=1}^{M} C_m/M$, the cost function of fuel consumption equilibrium is written as

$$\min J_4 = \sqrt{\sum_{m=1}^{M} \left(C_m - \overline{C}\right)^2} \tag{2.8}$$

Note that the above criteria are contradictory to some extent. Minimum fuel and minimum time may not be simultaneously satisfied. Therefore, spacecraft multi-mission planning is a multi-objective optimization problem (MOOP).

2.1.1.3 Constraints

1. Mission constraint
   Assume that each mission is accomplished by one servicer, then

$$\begin{cases} \sum\limits_{n=1}^{N} x_{mn} \leq a_m & m = 1, \ldots, M \\ \sum\limits_{m=1}^{M} x_{mn} \leq 1 & n = 1, \ldots, N \end{cases} \tag{2.9}$$

Considering the constraints of observation and control, defined by $[t^{\text{ocl}}, t^{\text{ocr}}]$, the maneuvering moment t_{mk}^{maneuver} of servicer m must be within the time window $[t_{mk}^{\text{ocl}}, t_{mk}^{\text{ocr}}]$, written as

$$\exists \left[t_{mk}^{\text{ocl}}, t_{mk}^{\text{ocr}}\right], t_{mk}^{\text{maneuver}} \in \left[t_{mk}^{\text{ocl}}, t_{mk}^{\text{ocr}}\right] \tag{2.10}$$

Defining Δt^{max} as the acceptable maximum far-range maneuver duration, the constraint of mission duration is expressed as

$$\max\{\Delta t_m\} \leq \Delta t^{\text{max}} \tag{2.11}$$

2. Resource constraint

Define $\Delta V_m^{\max}$ as the maximum maneuver capability offered by servicer m, then

$$\Delta V_m \le \Delta V_m^{\max} \tag{2.12}$$

Considering the duration requirement Δt_{ins} of command injection, the observation and control window $[t_{m0}^{\text{ocl}}, t_{m0}^{\text{ocr}}]$ before t_{m1}^{maneuver} should satisfy

$$\exists \left[t_{m0}^{\text{ocl}}, t_{m0}^{\text{ocr}}\right], \; t_{m0}^{\text{ocr}} - t_{m0}^{\text{ocl}} \ge \Delta t_{\text{ins}} \tag{2.13}$$

To sum up, a multilevel multi-objective nonlinear planning model is established as

$$
(P1)
\begin{cases}
\max J_1(\mathbf{X}) = \displaystyle\sum_{m=1}^{M}\sum_{n=1}^{N} x_{mn} p_n \\[2ex]
\min J_2(\mathbf{X}, \mathbf{Q}, \mathbf{T}) = \displaystyle\sum_{m=1}^{M}\left(\sum_{k=1}^{a_m} \Delta V_{mk}\right) \\[2ex]
\min J_3(\mathbf{X}, \mathbf{Q}, \mathbf{T}) = \max\left\{\displaystyle\sum_{k=1}^{a_m} \Delta t_{mk}\right\} \\[2ex]
\min J_4(\mathbf{X}, \mathbf{Q}, \mathbf{T}) = \sqrt{\displaystyle\sum_{m=1}^{M}(C_m - \overline{C})^2}
\end{cases}
\tag{2.14}
$$

$$
\text{s.t. } \sum_{n=1}^{N} x_{mn} \le a_m \quad m = 1, \dots, M
$$
$$
\sum_{m=1}^{M} x_{mn} \le 1 \quad n = 1, \dots, N
$$

where $\mathbf{Q}$ and $\mathbf{T}$ are the solutions to Eq. (2.15) for each given $\mathbf{X}$.

$$
(P2)
\begin{cases}
\min J_2(\mathbf{Q}, \mathbf{T}) = \displaystyle\sum_{m=1}^{M}(\Delta V_m) \\[2ex]
\min J_3(\mathbf{Q}, \mathbf{T}) = \max\{\Delta t_m\} \\[2ex]
\min J_4(\mathbf{Q}, \mathbf{T}) = \sqrt{\displaystyle\sum_{m=1}^{M}(C_m - \overline{C})^2}
\end{cases}
\tag{2.15}
$$

where $\mathbf{T}$ is the solution to Eq. (2.16) for each given $\mathbf{X}$ and $\mathbf{Q}$.

$$(P3) \begin{cases} \min \ J_2(\mathbf{T}) = \sum_{m=1}^{M} (\Delta V_m) \\ \min \ J_3(\mathbf{T}) = \max\{\Delta t_m\} \\ \min \ J_4(\mathbf{T}) = \sqrt{\sum_{m=1}^{M} (C_m - \overline{C})^2} \end{cases} \tag{2.16}$$

$$\text{s.t. } \Delta V_m \leq \Delta V_m^{\max}$$

$$\max\{\Delta t_m\} \leq \Delta t^{\max}$$

$$\exists \left[t_{mk}^{\text{ocl}}, t_{mk}^{\text{ocr}} \right], t_{mk}^{\text{maneuver}} \in \left[t_{mk}^{\text{ocl}}, t_{mk}^{\text{ocr}} \right]$$

$$\exists \left[t_{m0}^{\text{ocl}}, t_{m0}^{\text{ocr}} \right], t_{m0}^{\text{ocr}} - t_{m0}^{\text{ocl}} \geq \Delta t_{\text{ins}}$$

Note that Eqs. (2.14)–(2.16) are all MOOPs with conflicting interests, where the achievement of one objective may be at the cost of another and simultaneous optimality can hardly be secured. Trade-offs must be made among objectives to get Pareto-optimal solutions by multi-objective optimization algorithms.

Traditional multi-objective optimization approaches, such as weighted and constraint methods, transform an MOOP to several different single objective optimization problems (SOOP). However, these traditional approaches require separately solving SOOP many times, resulting in heavy computational burden.

Therefore, to ease the computation work by exploiting the correlation of different SOOPs, the Multi-Objective Evolutionary Algorithm (MOEA) is introduced and mostly applied to get Pareto-optimal solutions to an MOOP. Various MOEAs have flourished in literature such as Vector Evaluated Genetic Algorithm (VEGA) [2] and Niched Pareto Genetic Algorithm (NPGA) [3].

Considering the multilevel nature of Eqs. (2.14)–(2.16), it is difficult to generate the solution by MOEA directly. Therefore, we try to simplify the above model of Eqs. (2.14)–(2.16) by analyzing the spacecraft multi-mission properties, and then look for a proper solution strategy.

2.1.2 Solution Strategy

As mentioned above, the spacecraft multi-mission planning could be divided into mission assignment, sequence arrangement, and time distribution. Therefore, an embedded two-level solution scheme is designed. As illustrated in Fig. 2.2, the lower level is to deal with a two-level planning subsystem composed of sequence arrangement and time distribution, while the upper level is to deal with a two-level planning subsystem composed of mission assignment and the lower subsystem.

The above scheme involves multidimensional discrete/continuous variables, generating a large solution space. According to the orbital deployment and maneuver

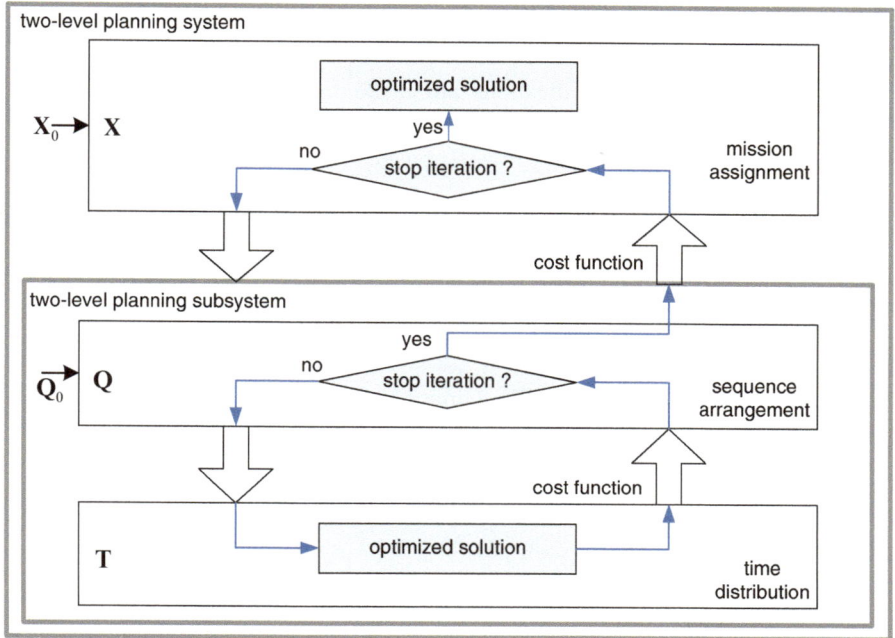

Fig. 2.2   Schematic of spacecraft multi-mission planning model

characteristics, we make proper problem-specific assumptions to explore efficient algorithms.

1. It is difficult for the servicer to make out-of-plane maneuvers due to its limited orbital maneuver capability. If the customers are scattered in different planes, or if the missions are time-demanding, one servicer can only execute one mission. In this way, the model is transformed to a mission assignment problem.
2. If the customers share the same orbit, or if the missions are not time-demanding, one servicer can be given several missions, thus transforming the model to a sequence arrangement and time distribution problem.

2.2   Integer Programming Method for Mission Assignment

Assuming that one servicer only serves one customer, the mission assignment problem is essentially to select a customer for each servicer, thus making sure those with higher priority are first addressed with less time and fuel, and better observation and control.

2.2.1 Planning Model

In Eqs. (2.14)–(2.16), simplifications are made by neglecting the lower-level sequence arrangement and time distribution, represented by $a_m = 1$, to derive the model of mission assignment as

$$
\begin{cases}
\max J_1(\mathbf{X}) = \sum_{m=1}^{M}\sum_{n=1}^{N} x_{mn}p_n \\[2mm]
\min J_2(\mathbf{X}) = \sum_{m=1}^{M} \Delta V_{m1} \\[2mm]
\min J_3(\mathbf{X}) = \max\{\Delta t_{m1}\} \\[2mm]
\min J_4(\mathbf{X}) = \sqrt{\sum_{m=1}^{M}(C_m - \overline{C})^2}
\end{cases}
$$

s.t.   (2.17)

$$
\sum_{n=1}^{N} x_{mn} \le a_m \quad m = 1, \ldots, M
$$

$$
\sum_{m=1}^{M} x_{mn} \le 1 \quad n = 1, \ldots, N
$$

$$
\Delta V_m \le \Delta V_m^{\max}
$$

$$
\max\{\Delta t_m\} \le \Delta t^{\max}
$$

$$
\exists \left[t_{m1}^{\mathrm{ocl}}, t_{m1}^{\mathrm{ocr}}\right], t_{m1}^{\mathrm{maneuver}} \in \left[t_{m1}^{\mathrm{ocl}}, t_{m1}^{\mathrm{ocr}}\right]
$$

$$
\exists \left[t_{m0}^{\mathrm{ocl}}, t_{m0}^{\mathrm{ocr}}\right], t_{m0}^{\mathrm{ocr}} - t_{m0}^{\mathrm{ocl}} \ge \Delta t_{\mathrm{ins}}
$$

which can be treated as an integer programming problem by separately dealing with maximum and minimum criteria as

$$
\begin{aligned}
\max J &= J_1 \\
\min J' &= \alpha_2 J_2 + \alpha_3 J_3 + \alpha_4 J_4
\end{aligned}
\tag{2.18}
$$

where the first cost function is maximum mission priority, and the second is minimum weighting combination of fuel, time, and fuel consumption equilibrium.

2.2.2 Algorithms

The above model can be solved by the available powerful Mixed Integer Linear Programming (MILP) solvers, such as CPLEX, EXPRESS, and many others. If the mission involves a small number of spacecraft, even the Enumeration

Table 2.1 Initial conditions in simulation

Spacecraft	a (km)	e	i (deg)	Ω (deg)	ω (deg)	f (deg)
C-1	7,250	0.03	94	218	10	15
C-2	7,300	0.01	95	215	20	50
C-3	7,250	0.03	104	220	25	230
C-4	7,300	0.02	103	222	0	130
S-1	7,150	0.01	98	220	30	50
S-2	7,150	0.01	98	220	30	230

Table 2.2 Simulation results

Optimized scheme	J	J_2	J_4
S-1 ~ C-3, S-2 ~ C-4	1.6	2183.4 m/s	1256.5 m/s

method is feasible. Here, we try to integrate stratification method and linear weighting method to transform Eq. (2.18) to a general integer programming model.

First, we find the optimized solutions to the first cost function in Eq. (2.18), thus constituting a set of all feasible solutions, referred to as S_0. Second, we optimize the second cost function in S_0. The detailed steps are given as follows.

Step 1: Order the customers by mission priority, and check whether the first M customers satisfy the constraints. If not, delete them in the customer set and fill the vacancy with subsequent candidates until M qualified customers are found or all customers are checked over. So far we have the optimized solution set to the first cost function.

Step 2: Take the above solution set as the customers and use the branch and bound algorithm to solve the second cost function in Eq. (2.18).

2.2.3 Numerical Simulation

Considering the actual circumstances in OOS, we take $M = 2, N = 4$, and $\Delta t^{\text{max}} = 5$ h. The initial orbital elements are given as in Table 2.1. The tracking, telemetering, and command (TT&C) ground stations include Xi'an, Xiamen, Neimeng, Kashi, Jiaodong, Beijing, Changchun, Weinan, Handan, Jinan, Qingdao, Taiyuan, Yuanwang-1 (longitude $70°$, latitude $0°$), Yuanwang-2 (longitude $-20°$, latitude $-20°$), Yuanwang-3 (longitude $-150°$, latitude $0°$). The velocity increment offered by the two servicers is $\{2,000\,\text{m/s} \quad 1,900\,\text{m/s}\}$, the initial moment is 2010-01-01 01:00:00 UTC, and the instruction injection lasts for 30 s, i.e., $\Delta t_{\text{ins}} = 30$ s.

Assume the mission priority of the four customers as $\{0.7 \quad 0.6 \quad 0.9 \quad 0.7\}$. The previous algorithm yields the results as shown in Table 2.2.

From Table 2.2, the optimized assignment scheme is Servicer-1 serving Customer-3, and Servicer-2 serving Customer-4. In the former, the servicer stays at the initial orbit for 800 s and then maneuvers; in the latter, the servicer stays at the initial orbit for 3,650 s and then maneuvers. Simulation results show that the derived scheme can offer a feasible optimized solution to address the mission assignment problem.

2.3   HGABB for One-to-N Spacecraft Mission Planning

The one-to-N spacecraft mission planning problem is primarily concerned with sequence arrangement and time distribution. The former aims to seek the best serving sequence and the latter seeks the time distribution that is assigned to each mission, such that the cost function is minimized. The basic solution procedures could be described as below.

Step 1: Select an initial mission sequence.
Step 2: Optimize the time distribution for this mission sequence.
Step 3: Produce a new mission sequence by an optimization algorithm.
Step 4: Iterate Steps 2 and 3 until an acceptable mission sequence and corresponding time distribution are found.

In this sense, one-to-N spacecraft mission planning is a two-level optimization problem, with the leader representing sequence arrangement and the subordinate representing time distribution. Given the satellite density in GEO orbit, the one-to-N approach helps save cost and boost efficiency. In the following section, the analysis is carried out against the inspection scenario over multiple GEO satellites [4, 5], as shown in Fig. 2.3; then the one-to-N servicing mission could be implemented by coplanar phase adjustment of the servicer.

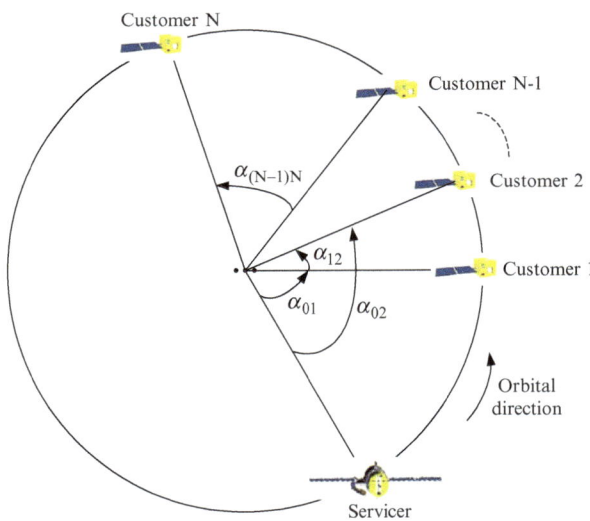

Fig. 2.3 Schematic of one-to-N servicing mission

2.3.1 Planning Model

According to Sect. 2.1.1 and the second assumption in Sect. 2.1.2, the one-to-N spacecraft mission planning model could be simplified as a two-level planning model

$$(P1) \ \min J_2(\mathbf{Q}, \mathbf{T}) = \sum_j \sum_i \Delta V_{ij} q_{ij}$$

$$\text{s.t.}$$
$$\sum_i q_{ij} = 1$$
$$\sum_j q_{ij} = 1 \qquad (2.19)$$
$$\sum_j \sum_i \Delta V_{ij} q_{ij} \leq \Delta V^{max}$$

where $i, j \in [1, \ldots, N]$ are the indices of customers of two adjacent missions, and q_{ij} the element of $\mathbf{Q}$ which satisfy

$$q_{ij} = \begin{cases} 1 & \text{maneuver to the } j^{th} \text{ customer after serving the } i^{th} \\ 0 & \text{otherwise} \end{cases} \qquad (2.20)$$

and ΔV_{ij} is the velocity increment needed to maneuver from the ith to jth customer which satisfies $\Delta V_{ii} = 0$ and $\Delta V_{ij} \neq \Delta V_{ji}$. ΔV_{ij} depends on α_{ij} and Δt_{ij}^Q and could be solved by the two-impulse multiple-revolution Lambert transfer algorithm (see Sect. 3.1.1), and Δt_{ij}^Q is the time spent by the servicer to maneuver from the ith to jth customer in a mission sequence $\mathbf{Q}$.

The decision variable vector $\mathbf{T}$ could be solved from the lower-level model which is derived to optimize the time distribution, written as

$$(P2) \ \min J_2(\mathbf{T}) = \sum_j \sum_i \Delta V_{ij}\left(\alpha_{ij}, \Delta t_{ij}^Q\right)$$

$$\text{s.t.} \sum_j \sum_i \Delta V_{ij}\left(\alpha_{ij}, \Delta t_{ij}^Q\right) \leq \Delta V^{max}$$
$$\sum_j \sum_i \Delta t_{ij}^Q \leq \Delta T^{max} \qquad (2.21)$$
$$\Delta V_{ij}\left(\alpha_{ij}, \Delta t_{ij}^Q\right) \leq \kappa \Delta V^{max}$$
$$\Delta t_{ij}^Q \leq \rho \Delta T^{max}$$

where κ and $\rho(0 < \rho, \kappa < 1)$ are proportional coefficients, which pose restrictions on reasonable cost. And, $\alpha_{ij} \in [-\pi, \pi]$ is the phase difference between the ith to jth GEO satellite.

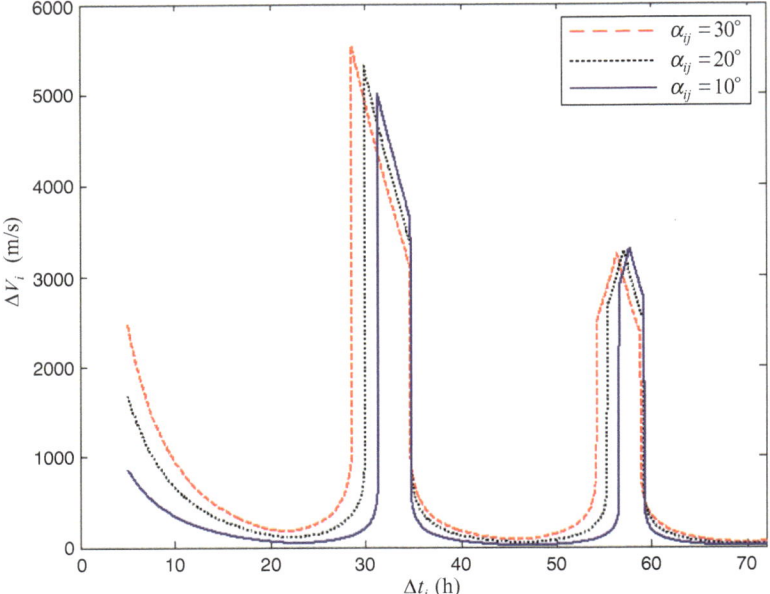

Fig. 2.4 Relationship between Δt_i and ΔV_i at GEO

2.3.2 Algorithms

The key to solving Eq. (2.23) is the dynamics property, i.e., the relationship between one certain mission duration Δt_i and velocity increment ΔV_i, which can be addressed by fixed-time two-impulse multiple-revolution Lambert solution (see Sect. 3.1.1) [1]. For a GEO scenario, consider the rendezvous of two co-orbital spacecraft while the maneuvering moment is chosen at the initial time. The result is shown in Fig. 2.4, and we can conclude that ΔV_{ij} can be optimized by adjusting the first impulse moment.

The optimized relationship is shown in Fig. 2.5 which indicates that ΔV_i keeps decreasing with Δt_i. The curves can be viewed as comprising stabilizing and descending segments, which occur in turn. The interval of stabilizing segment, which is usually long, depends on the interval between two local minimum values in Fig. 2.4. Except for the first descending segment, others are of short time. And in descending segments, the velocity increment is time-sensitive. Therefore, simplifications are made by limiting Δt_i in the stabilizing segment. In this way, the problem of time distribution is summarized as how to choose the Δt_i-located segment and determine its specific value.

Define s_i^{num} as the number of the stabilizing segment in Δt_i and $\Delta V_{is}(s = 1, 2, \cdots, s_i^{\text{num}})$ as the velocity increment in the sth stabilizing segment. Further, define t_{is}^{start} as the start moment of the sth stabilizing segment and t_{is}^{end} as its end moment

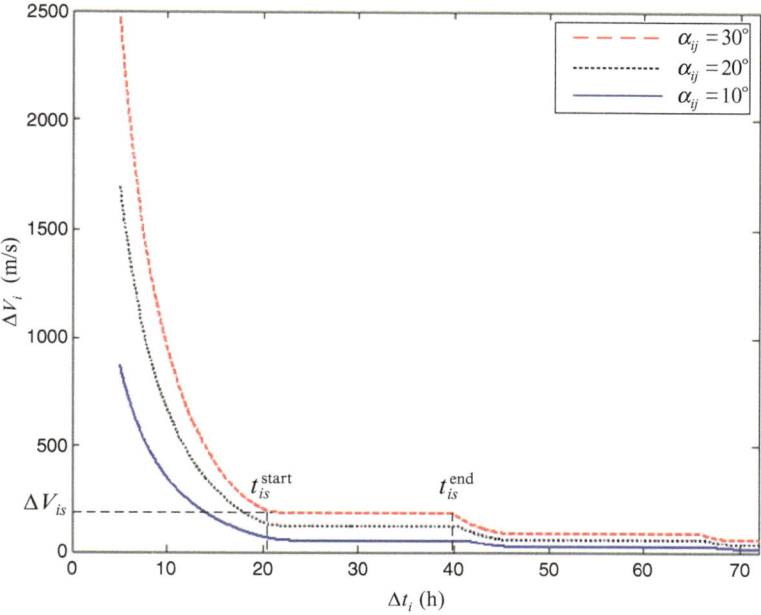

Fig. 2.5 Optimized relationship between Δt_i and ΔV_i at GEO

with respect to the ith mission initial moment. In this way, the lower planning model is transformed to a 0–1 integer programming problem, written as

$$\min \sum_{i=1}^{N} \sum_{s=1}^{s_i^{\text{num}}} \Delta V_{is} y_{is}$$

$$\text{s.t.} \quad y_{is} = \begin{cases} 1 & \Delta t_i \in \left[t_{is}^{\text{start}}, t_{is}^{\text{end}} \right], j = 1, 2, \ldots, s_i^{\text{num}} \\ 0 & \text{otherwise} \end{cases}$$

$$\sum_{s=1}^{s_i^{\text{num}}} y_{is} = 1$$

$$\sum_{i=1}^{N} \sum_{s=1}^{s_i^{\text{num}}} t_{is}^{\text{end}} y_{is} \geq \Delta T^{\max} \tag{2.22}$$

$$\sum_{i=1}^{N} \sum_{s=1}^{s_i^{\text{num}}} t_{is}^{\text{start}} y_{is} \leq \Delta T^{\max}$$

$$\sum_{i=1}^{N} \sum_{s=1}^{s_i^{\text{num}}} \Delta V_{is} y_{is} \leq \Delta V^{\max}$$

The first step to solve this model is to calculate t_{is}^{start}, t_{is}^{end}, and ΔV_{is} by the two-impulse multiple-revolution Lambert solution. Then, a 0–1 integer programming

Fig. 2.6 HGABB
algorithm

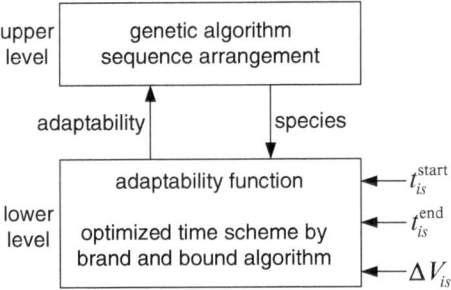

Table 2.3 Simulation results of case 1

Planning algorithm	Enumeration	HGABB
Optimized sequence arrangement	3-2-1-4-5-6	3-2-1-4-5-6
Optimized time distribution (h)	[23.81, 47.92, 23.81, 71.14, 23.61, 47.31]	[23.81, 47.92, 23.81, 71.14, 23.61, 47.31]
Velocity increment (m/s)	68.61	68.61

problem can be solved by common methods of enumeration or branch and bound, the computation amount of which is highly dependent on Δt_i and N. If N is small, enumeration method is better due to its capability of generating a global optimal solution. ΔV_{is} is determined by the Δt_i-located segment and Δt_i can be at any point in the segment. If N is big, the enumeration method is not feasible. Instead, we use the genetic algorithm [6].

The algorithm of this two-level nonlinear planning model is given in Fig. 2.6.

2.3.3   Numerical Simulation

Case 1: Take a six GEO satellites inspection mission for example. Assume the mission duration is 240 h, and the initial phase relationship is expressed as

$$[\alpha_{01}, \alpha_{12}, \alpha_{23}, \alpha_{34}, \alpha_{45}, \alpha_{56}] = [-10°, 1°, 3°, 5°, 2°, 7°] \tag{2.23}$$

Simulation results are derived by enumeration and HGABB respectively, given as in Table 2.3, which verify the effectiveness of the latter.

Case 2: Take a ten GEO satellites inspection mission for example. Assume the mission duration is 360 h and 720 h respectively, and the initial phase relationship is expressed as

$$\begin{aligned} &[\alpha_{01} \quad \alpha_{12} \quad \alpha_{23} \quad \alpha_{34} \quad \alpha_{45} \quad \alpha_{56} \quad \alpha_{67} \quad \alpha_{78} \quad \alpha_{89} \quad \alpha_{910}] \\ &= [37° \quad 10° \quad 13° \quad 4° \quad 32° \quad 5° \quad 6° \quad 12° \quad 25° \quad 7°] \end{aligned} \tag{2.24}$$

Table 2.4 Simulation results of case 2

Mission duration	360 h (both-optimal)	720 h (both-optimal)
Mission sequence	4-3-2-1-5-6-7-8-9-10	1-2-3-4-5-6-7-8-9-10
Time distribution (h)	24.42, 24.00, 48.64, 24.42, 67.89, 23.39, 23.33, 46.97, 46.11, 23.25	48.44, 72.00, 48.64, 48.44, 67.89, 71.42, 71.33, 70.94, 70.08, 71.28
Velocity increment (m/s)	500.4	320.4
$\sum \|\alpha_{ij}\|$	151°	151°
min $\sum \|\alpha_{ij}\|$	151°	151°

Table 2.5 Comparison of different scenarios

Mission duration	720 h (both-optimal)	720 h (sequence-optimal)	720 h (time-optimal)
Mission sequence	1-2-3-4-5-6-7-8-9-10	1-2-3-4-5-6-7-8-9-10	5-6-7-8-9-10-4-3-2-1 (given)
Time distribution (h)	48.44, 72.00, 48.64, 48.44, 67.89, 71.42, 71.33, 70.94, 70.08, 71.28	Mean	70.28, 71.42, 71.33, 70.94, 70.08, 71.28, 53.58, 72.00, 48.64, 48.44
Velocity increment (m/s)	320.4	445.4	441.7
$\sum \|\alpha_{ij}\|$	151°	151°	182°
min $\sum \|\alpha_{ij}\|$	151°	151°	151°

HGABB generates the planning results as given in Table 2.4. A comparison reveals that mission sequence arrangement and fuel consumption differ as mission duration changes. In Table 2.4, min $\sum \|\alpha_{ij}\|$ is the minimum value of $\sum \|\alpha_{ij}\|$ in all feasible sequences. Simulation results show that $\sum \|\alpha_{ij}\|$ and min $\sum \|\alpha_{ij}\|$ share the same value, indicating that $\sum \|\alpha_{ij}\|$ in the optimal sequence usually takes the minimum value, and phase relationship is key to fuel consumption. Table 2.5 lists the results of both-optimal, sequence-optimal, and time-optimal scenarios with 720 h mission duration. Simulation results show that fuel consumption is greatly reduced in both-optimal scenario.

References

1. Shen, H. J. (2003). *Optimal scheduling for satellite refuelling in circular orbits*. Dissertation, Georgia Institute of Technology.
2. Schaffer, J. D. (1985). *Multiple objective optimization with Vector Evaluated Genetic algorithms*. Dissertation, Vanderbilt University.
3. Horn, J., Nafpliotis, N., & Goldberg, D. E. (1994). *A niched Pareto genetic algorithm for multiobjective optimization*. IEEE World Congress on Computational Computation.

4. Alfriend, K. T., Lee, D. J., & Clenn, C. N. (2002). *Optimal servicing of geosynchronous satellites*. AIAA/AAS Astrodynamics Specialist Conference and Exhibit, Monterey, CA.
5. Dutta, A., & Tsiotras, P. (2008). An egalitarian peer-to-peer satellite refueling strategy. *AIAA Journal of Spacecraft and Rockets, 45*(3), 608–618.
6. Wang, G. M. (2004). *Two-level planning algorithm based on genetic algorithm*. MS Thesis, Wuhan University.

Chapter 3
Far-Range Orbital Maneuver Planning

Abstract Far-range orbital maneuver is an important on-orbit operations phase during which TT&C provides measurement and tracking of both the servicer and customer until the former gets close to the latter. Then the servicer sensor captures the customer, which marks the beginning of proximity relative motion phase. Trajectory planning of far-range orbital maneuver is to optimize the orbital maneuver for the servicer under the constraints of time, fuel, and observation and control. The investigation on orbital maneuver generally employs impulse thrust method and finite continuous thrust method. Considering the advantage of the former in simplifying the problem and producing approximate results in preliminary design and demonstration, this chapter is mainly based on impulse thrust. On the basis of Lambert solution, the algorithms of GA and randomized A* tree expansion are adopted to address the multi-impulse orbital maneuver optimization problem respectively.

3.1 Problem Formulation

A two-body problem is assumed and the customer is free. The dynamic model is formulated in the Earth Centered Inertia (ECI) reference frame, whose origin is located at the Earth's center, X axis points to the vernal equinox, Z axis points to the North Pole perpendicular to the equatorial plane, and Y axis is determined by the right-hand rule.

Denote the subscripts c and s as the customer and servicer respectively, $\mathbf{R}$ the absolute position vector, then

$$\ddot{\mathbf{R}}_s = -\frac{\mu \mathbf{R}_s}{R_s{}^3} + \mathbf{f}_{ds} + \frac{\mathbf{F}_s}{m_s} \qquad (3.1)$$

L. Yang et al., *On-Orbit Operations Optimization: Modeling and Algorithms*, SpringerBriefs in Optimization, DOI 10.1007/978-1-4939-0838-7_3, © Leping Yang, Yanwei Zhu, Xianhai Ren, Yuanwen Zhang 2014

Fig. 3.1 Lambert transfer
orbit with a high
eccentricity

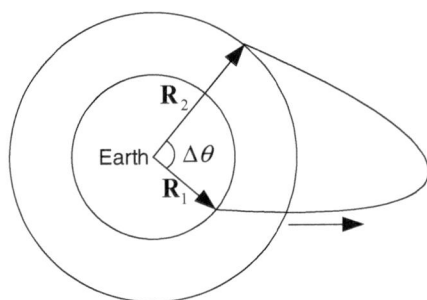

$$\ddot{\mathbf{R}}_c = -\frac{\mu \mathbf{R}_c}{R_c{}^3} + \mathbf{f}_{dc} \tag{3.2}$$

where μ is Earth gravitational constant, $\mathbf{f}_d$ perturbation acceleration offered by all perturbation forces acting on the spacecraft except for the gravity, $\mathbf{F}_s$ control force acting on the servicer, and m_s mass of the servicer.

3.1.1 Lambert Solution

The problem of two-impulse Lambert rendezvous [1] can be described as follows. Let the superscripts "$-$" and "$+$" denote the moment just before and after the ignition, and given the servicer positions $\mathbf{R}_1$, $\mathbf{R}_2$ and velocities $\mathbf{V}_1^-$, $\mathbf{V}_2^+$ at the two ignition moments t_1, $t_2(t_2 > t_1)$ respectively, and the transfer duration $\Delta T = t_2 - t_1$ of the servicer from $\mathbf{R}_1$ to $\mathbf{R}_2$, determine the velocity increment $\Delta \mathbf{V}_1$ and $\mathbf{V}_2^-$. Then $\Delta \mathbf{V}_2 = \mathbf{V}_2^+ - \mathbf{V}_2^-$. Note that the customer runs to $\mathbf{R}_2$ at t_2.

If ΔT is small, there exists a one and only Lambert transfer orbit, which can be solved by Gauss method, p-iteration method, and so on. However, If ΔT is big, the transfer orbit obtained by the above methods would have a high eccentricity, causing mounting fuel consumption, as shown in Fig. 3.1.

Therefore, we try to find such a transfer orbit in which the servicer runs N revolutions, then gets to the customer position, as shown in Fig. 3.2.

Given $\mathbf{R}_1$, $\mathbf{R}_2$, ΔT and the initial maneuver direction of the servicer, $\mathbf{V}_1^+$ and $\mathbf{V}_2^-$ can be solved by introducing the universal variable z as

$$\mathbf{V}_1^+ = \frac{\mathbf{R}_2 - f\mathbf{R}_1}{g}, \quad \mathbf{V}_2^- = \frac{\dot{g}\mathbf{R}_2 - \mathbf{R}_1}{g} \tag{3.3}$$

$$f = 1 - \frac{y(z)}{R_1}, \quad g = A\sqrt{\frac{y(z)}{\mu}}, \quad \dot{g} = 1 - \frac{y(z)}{R_2} \tag{3.4}$$

Fig. 3.2 *N*-revolution
Lambert transfer orbit

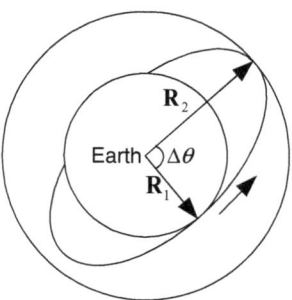

$$C(z) = \begin{cases} \dfrac{1 - \cos\sqrt{z}}{z} & \text{if } z > 0 \\[2ex] \dfrac{1}{2!} - \dfrac{z}{4!} + \dfrac{z^2}{6!} + \cdots = \sum_{k=0}^{\infty} \dfrac{(-z)^k}{(2k+2)!} & \text{if } z \to 0 \\[2ex] \dfrac{1 - \cosh\sqrt{-z}}{z} & \text{if } z < 0 \end{cases} \tag{3.5}$$

$$S(z) = \begin{cases} \dfrac{\sqrt{z} - \sin\sqrt{z}}{\sqrt{z^3}} & \text{if } z > 0 \\[2ex] \dfrac{1}{3!} - \dfrac{z}{5!} + \dfrac{z^2}{7!} + \cdots = \sum_{k=0}^{\infty} \dfrac{(-z)^k}{(2k+3)!} & \text{if } z \to 0 \\[2ex] \sinh\sqrt{-z} - \sqrt{-z} & \text{if } z < 0 \end{cases} \tag{3.6}$$

$$\Delta T = \frac{1}{\sqrt{\mu}} \left(x^3(z)S(z) + A\sqrt{y(z)} \right) \tag{3.7}$$

$$x(z) = \sqrt{\frac{y(z)}{C(z)}}, \quad y(z) = R_1 + R_2 - A\frac{(1 - zS(z))}{\sqrt{C(z)}}, \quad A = \frac{\sqrt{R_1 R_2} \sin \Delta\theta}{\sqrt{1 - \cos \Delta\theta}} \tag{3.8}$$

where $\Delta\theta$ represents the angle between $\mathbf{R}_1$ and $\mathbf{R}_2$, determined by the maneuver direction.

The total velocity increment is

$$\Delta V = \|\Delta \mathbf{V}_1\| + \|\Delta \mathbf{V}_2\| \tag{3.9}$$

Solving z in Eq. (3.7) is critical to the Lambert problem. Define

$$t(z) = \frac{1}{\sqrt{\mu}} \left(x^3(z)S(z) + A\sqrt{y(z)} \right) \tag{3.10}$$

then the relationship between z and $t(z)$ is indicated as in Fig. 3.3.

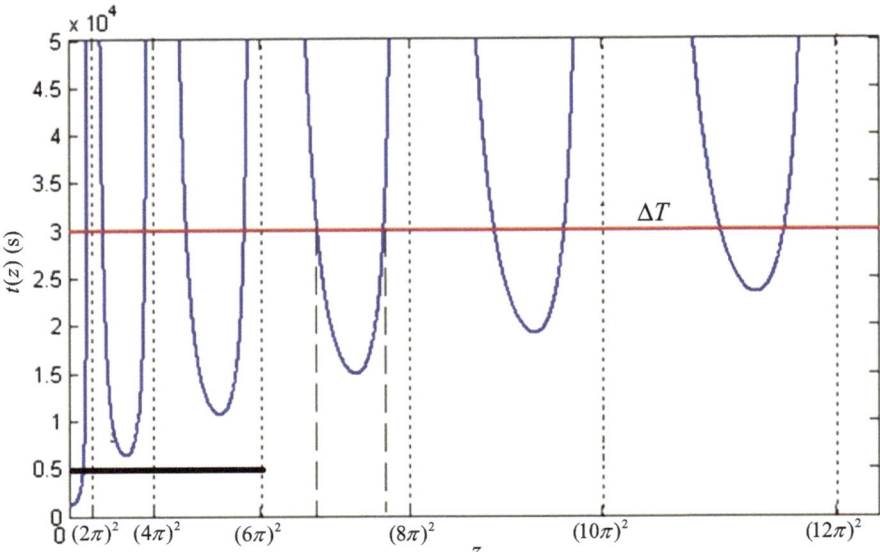

Fig. 3.3 $t(z)$-z relationship

Figure 3.3 shows the curves in $[(2n\pi)^2, (2(n+1)\pi)^2](n = 1, 2, \ldots, N)$ have similar periodicity. If ΔT is small, the black line of $t(z) = \Delta T$ and the curve have only one intersection in $[0, (2\pi)^2]$. By contrast, if ΔT is big, the red line of $t(z) = \Delta T$ and the curve have more than one intersection, corresponding to the possibility of several Lambert transfer orbits. Therefore, Lambert solution can be obtained by computing z at every intersection.

In Fig. 3.3, the curve in $[(2n\pi)^2, (2(n+1)\pi)^2]$ has a one and only local minimum t_n^{min} whose corresponding z is denoted as z_n^{min}. When $t_n^{min} > \Delta T$, the line of $t(z) = \Delta T$ has no intersection with the curve, indicating no transfer orbit in $[(2n\pi)^2, (2(n+1)\pi)^2]$. When $t_n^{min} < \Delta T$, the line of $t(z) = \Delta T$ has two intersections with the curve, indicating two transfer orbits. When $t_n^{min} = \Delta T$, the line of $t(z) = \Delta T$ has a one and only intersection with the curve, indicating only one transfer orbit.

After generating the transfer orbit, we need to justify its feasibility by guaranteeing the orbital altitude is always higher than 100 km (approximate atmospheric altitude).

To sum up, a scheme to solve N-revolution Lambert transfer problem is given, shown as in Fig. 3.4.

3.1.2 Multi-Impulse Trajectory Planning Model

Define a mission scenario as the servicer rendezvous with the customer by N impulse maneuvers. The mission starts at t_0 and finishes before t_f, as depicted in Fig. 3.5.

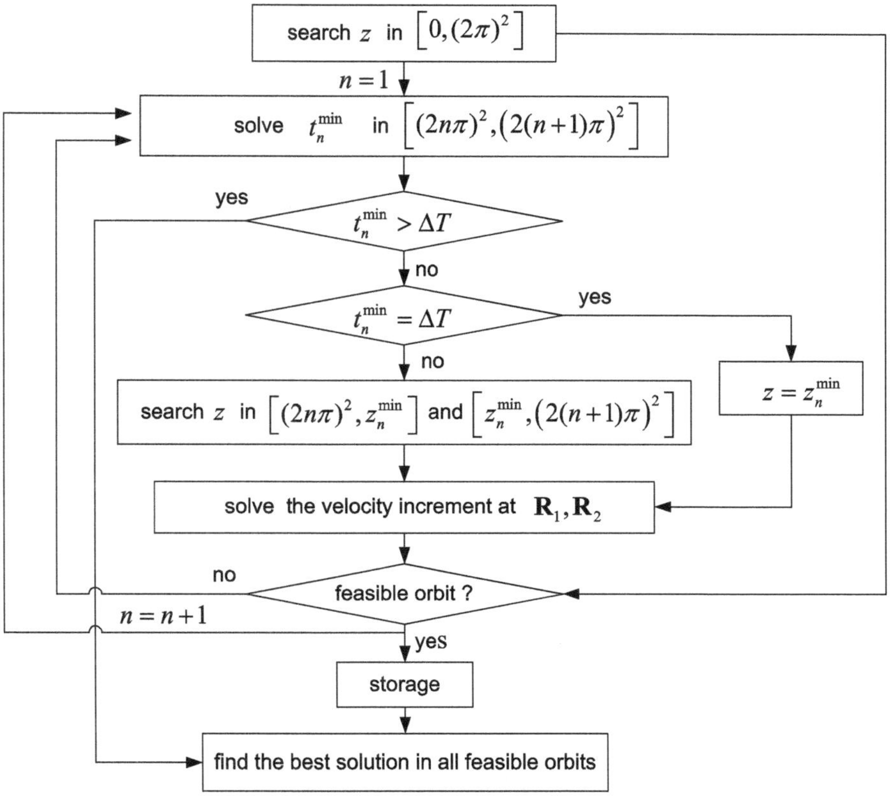

search z in $\left[0,(2\pi)^2\right]$
$n=1$
solve $t_n^{\min}$ in $\left[(2n\pi)^2,(2(n+1)\pi)^2\right]$
yes
$t_n^{\min} > \Delta T$
no
yes
$t_n^{\min} = \Delta T$
no
search z in $\left[(2n\pi)^2, z_n^{\min}\right]$ and $\left[z_n^{\min},(2(n+1)\pi)^2\right]$
$z = z_n^{\min}$
solve the velocity increment at $\mathbf{R}_1, \mathbf{R}_2$
no
feasible orbit ?
$n = n+1$
yes
storage
find the best solution in all feasible orbits

Fig. 3.4 Schematic of N-revolution Lambert transfer algorithm

t_N, $\mathbf{R}_N$, $\mathbf{V}_N^- + \Delta\mathbf{V}_N$
Customer orbit
t_f
t_0
t_i, $\mathbf{R}_i$, $\mathbf{V}_i^- + \Delta\mathbf{V}_i$
t_2, $\mathbf{R}_2$, $\mathbf{V}_2^- + \Delta\mathbf{V}_2$
t_f
t_0
t_1, $\mathbf{R}_1$, $\mathbf{V}_1^- + \Delta\mathbf{V}_1$
Servicer initial orbit

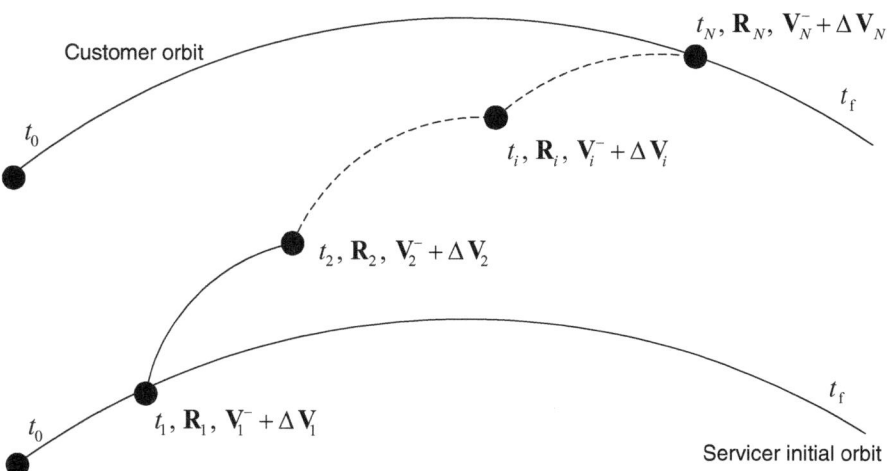

Fig. 3.5 Far-range orbital maneuver strategy

Define t_i ($i = 1, \ldots, N$) as the ith ignition moment of the servicer, $\mathbf{R}_i$, $\mathbf{V}_i^-$, and $\Delta\mathbf{V}_i$ as the corresponding position, velocity, and velocity increment. Note that the customer runs to $\mathbf{R}_N$ at t_N, with a velocity of $\mathbf{V}_N^- + \Delta\mathbf{V}_N$.

The multi-impulse trajectory planning model of far-range orbital maneuver is given as follows.

1. Decision variables
   Choose $2N$ decision variables: $t_1, \ldots, t_N$ and $\Delta\mathbf{V}_1, \ldots, \Delta\mathbf{V}_N$.
2. Cost function
   According to specific mission requirements, the cost function can be chosen as any one in

$$
\begin{aligned}
&\min \sum_{i=1}^{N} \|\Delta\mathbf{V}_i\| \\
&\min\ (t_N - t_0) \\
&\min \eta(t_N - t_0)/\Gamma + (1 - \eta)\sum_{i=1}^{N} \|\Delta\mathbf{V}_i\|
\end{aligned}
\tag{3.11}
$$

where $\eta \in (0, 1)$ is weighting ratio and Γ is a user-defined constant which denotes the tolerable mission duration increase corresponding to 1 m/s velocity increment decrease.

3. Constraints
   The terminal state constraint is expressed as

$$
\mathbf{R}_N = \mathbf{R}_\mathrm{d}, \mathbf{V}_N^+ = \mathbf{V}_\mathrm{d} = \mathbf{V}_N^- + \Delta\mathbf{V}_N
\tag{3.12}
$$

where $\mathbf{R}_\mathrm{d}$ and $\mathbf{V}_\mathrm{d}$ are the desired terminal state determined by the customer orbit. The observation and control constraint requires every ignition be in the observation and control window.

3.2   Genetic Algorithm for Multi-Impulse Planning

Intelligent optimization algorithms represented by Genetic Algorithm (GA) are characterized by high robustness, good global convergence, and parallelism, which make it advantageous in solving complicated optimization problems, especially in finding the global optimal solution. Given N, GA is a better choice to solve the model in Sect. 3.1.2.

3.2.1   Genetic Algorithm

GA [2, 3] is one of the intelligent optimization algorithms built on the basis of natural selection and genetic mechanism in the biological world. It adopts population search strategy and information exchange among members. The search of GA is independent of gradient information, which makes it superior to conventional approaches in addressing complicated nonlinear problems.

GA has four main features. First, the calculation is done against the coding group rather than the parameters. Second, the search is executed from the coding group rather than an individual. Third, the search is based on the cost function (fitness), and does not require such information as derivatives. Fourth, the operators of selection, crossover, and mutation are random operations.

The steps of GA generally include:

1. Coding. Before the search, GA encodes the data in the solution space with gene strings in the genetic space. The combination of these strings constitutes different species.
2. Generation of initial population of individuals. N initial strings are produced, each of which constitutes an individual, N of which constitute a population. GA executes the iteration with these N strings as initial species.
3. Checking individual feasibility. Reject or repair those unqualified offspring.
4. Calculation of fitness. Fitness characterizes the superiority of individuals or solutions. Note that the definition of fitness function differs with problems.
5. Selection. This operation chooses some offspring for survival according to predefined rules. This keeps the population size within a fixed constant and puts good offspring into the next generation with a high probability.
6. Crossover. It generates offspring from two chosen individuals in the population, by exchanging some bits in the two individuals. The offspring thus inherit some characteristics from each parent. Crossover embodies the concept of information exchange.
7. Mutation. It generates offspring by randomly changing one or several bits in an individual. Offspring may thus possess different characteristics from their parents. Mutation provides an opportunity for the production of new individuals.

The basic procedures of GA are illustrated in Fig. 3.6.

3.2.2 Planning Model

As depicted in Fig. 3.7, given $t_3, \ldots, t_N$ and $\Delta V_3, \ldots, \Delta V_N$, we can derive $\mathbf{R}_2$ and $\mathbf{V}_2^+$ based on the terminal constraints. Meanwhile, according to $\mathbf{R}_0$ and $\mathbf{V}_0$, we can derive $\mathbf{R}_1$ and $\mathbf{V}_1^-$. Then, the first two impulses can be solved by Lambert solution. The problem is summarized as looking for the optimal solution to $t_1, \ldots, t_N$ and $\Delta V_1, \ldots, \Delta V_N$.

Introducing the variable $\alpha_i \in (0, 1)$ $(i = 1, \ldots, N)$, the ignition moments t_i can be written as

$$\begin{cases} t_i = t_0 + \alpha_i(t_{i+1} - t_0) & i = 1, \ldots, N - 1 \\ t_N = t_0 + \alpha_N(t_f - t_0) \end{cases} \tag{3.13}$$

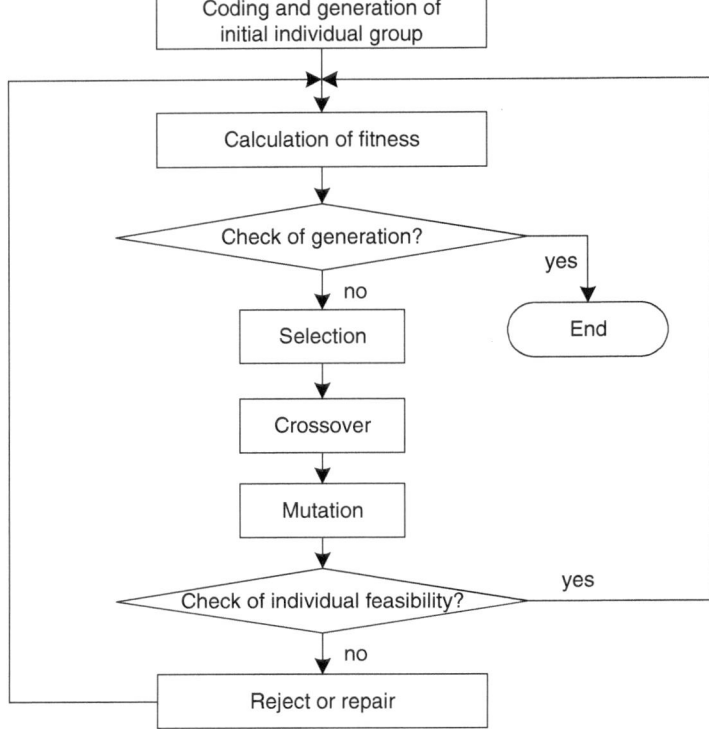

Fig. 3.6 GA procedures

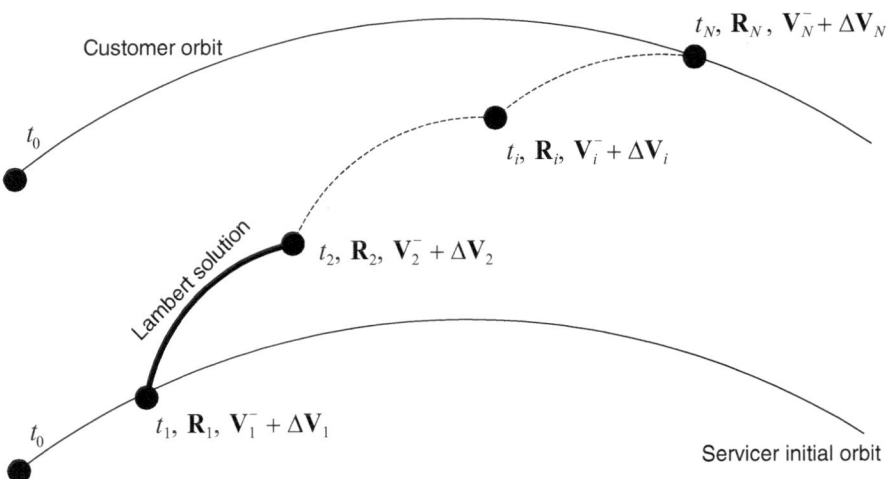

Fig. 3.7 Multi-impulse trajectory planning via GA

In this way, the decision variables of $t_1, \ldots, t_N$ are transcribed to α_i, which has the advantage of limiting the variables in $(0,1)$, convenient for GA search.

Define $\Delta V_{\max}$ as the maximum velocity increment of each ignition, $\pi/2$ as the maximum angle between the impulse direction and x-axis (which corresponds to the Fig. 3.7 scenario, if the servicer is higher than the customer, then it's x-axis) in the orbital reference frame. Assume all impulse vectors are coplanar with the customer orbit excluding the first two impulses. Introducing the variables $\beta_i \in (0, 1)$ $(i = 1, \ldots, N)$ gives ΔV_i as

$$\Delta V_i = \beta_i \Delta V_{\max} \tag{3.14}$$

and introducing the variables $\gamma_i \in (0, 1)$ $(i = 1, \ldots, N)$ to give $\gamma_i \pi/2$ as the angle between $\Delta \mathbf{V}_i$ and x-axis.

Optimal fuel consumption is written as

$$J_1 = \min \sum_{i=1}^{N} |\Delta \mathbf{V}_i| \tag{3.15}$$

Optimal time consumption is written as

$$J_2 = \min (t_N - t_0) \tag{3.16}$$

Optimal weighting of fuel and time is written as

$$J_3 = \min \left((1 - \eta) \sum_{i=1}^{N} |\Delta \mathbf{V}_i| + \eta(t_N - t_0)/\Gamma \right) \tag{3.17}$$

For the convenience of GA search, we further transform the cost functions to

$$\max J' = -\left(J + \sum_{j=1}^{M} S_j \right) \tag{3.18}$$

where M is the number of constraints and

$$J = \begin{cases} (t_N - t_0)/(t_f - t_0) & \text{time-optimal} \\ \sum_{i=1}^{N} (\Delta V_i/\Delta V_{\max}) & \text{fuel-optimal} \\ (1 - \eta) \sum_{i=1}^{N} (\Delta V_i/\Delta V_{\max}) + \eta(t_N - t_0)/(t_f - t_0) & \text{weighting-optimal} \end{cases} \tag{3.19}$$

S_j, which indicates whether the jth constraint is satisfied, is defined as

Table 3.1 Initial conditions in simulation

Spacecraft	a (km)	e	i (deg)	Ω (deg)	ω (deg)	f (deg)
Customer	7,430	0	80	220	0	120
Servicer	6,850	0	80	220	0	0

Table 3.2 Simulation results via GA

–	t (s)	ΔV (m/s)	ΔV_X (m/s)	ΔV_Y (m/s)	ΔV_Z (m/s)
Impulse-1	3,188	148.24	−30.15	12.60	−144.60
Impulse-2	23,090	207.32	55.72	34.24	196.74
Impulse-3	44,222	42.68	18.05	38.60	−2.28

$$S_j = \begin{cases} 0 & j\text{th constraint satisfied} \\ 1 & \text{otherwise} \end{cases} \qquad (3.20)$$

The GA initial population is randomly produced with some satisfying the constraint of observation and control and some not. We define the weighting threshold of the latter species as 1, and that of the former as 0. As a result, for those schemes violating observation and control constraint, the cost function J' would always be smaller than −1. So unqualified species have very little possibility to be chosen in GA's evolutionary computation, and usually end up with elimination. It is safely concluded the optimized solution generated by GA is sure to satisfy observation and control constraint.

3.2.3 Numerical Simulation

The initial orbital elements at $t_0 = 0$ are listed in Table 3.1. The mission requires the servicer maneuver to 30 km behind the customer within 24 h. The cost function is fuel-optimal. The observation and control constraint is same as the case in Sect. 2.2.3.

Suppose $N = 3$, $\Delta V_{\max} = 300$ m/s, and the number of generation and species are both 30. With different initial populations, we have carried out multiple simulations, one of whose results is given in Table 3.2.

The simulation shows that GA works well in addressing spacecraft far-range multi-impulse trajectory planning problems, but the results are suboptimal and random subjected to the choice of initial value.

3.3 Random Optimization for Multi-Impulse Planning

The randomized A* tree expansion algorithm [4, 5], based on randomized tree-based exploration techniques and A* network searches, is convenient in addressing various constraints (especially in optimizing the number of impulses)

Fig. 3.8 Pseudocode
of randomized A* tree
expansion algorithm

Algorithm
1   for $i = 0$ to *Num* do
2       $p = $ CHOOSE_WAYPOINT $()$
3       $n = $ EXPAND_WAYPOINT (p)
4   ADD_TO_TREE (p, n)
5   end for

and high in planning efficiency. Furthermore, compared with GA for a fixed N, the randomized A* tree expansion algorithm could take N as an additional decision variable.

3.3.1   Randomized A* Tree Expansion Algorithm

The pseudocode of the randomized A* tree expansion algorithm is given in Fig. 3.8.

The inputs to the algorithm are the initial and terminal states. The tree is seeded with the initial state and then built by applying the CHOOSE_WAYPOINT method to choosing a waypoint in the tree, which is then expanded to an appropriate new waypoint via the EXPAND_WAYPOINT method, and then the new waypoint is added to the tree, if possible, using the ADD_TO_TREE method which also attempts to connect the new waypoint to the terminal state. These procedures keep running when iteration terminates at a point where it returns to the best path, indicating a best solution. Note that iteration times are not necessarily related to the degree of optimization, but more iteration may result in better solutions.

The program terminates when *Num* iterations are run, or a feasible solution has been found. Experiments have shown that halting the iteration when a feasible solution is found returns a path quickly, but running for *Num* iterations may find a better transfer path when more search time is available.

The CHOOSE_WAYPOINT method determines which waypoint will be expanded by weighted distribution. We compute a weight for each waypoint and select a waypoint randomly with a probability proportional to its weight. This can be effectively and quickly done via a heap data structure.

The weight is calculated as

$$weight = \frac{f_1(order)}{f_2(A^*\_cost) \cdot f_3(out\_degree)} \qquad (3.21)$$

where the function f_i is incremental, and *order* represents the kinship of the point with the initial point. The value of *order* for the initial point is 1; the value

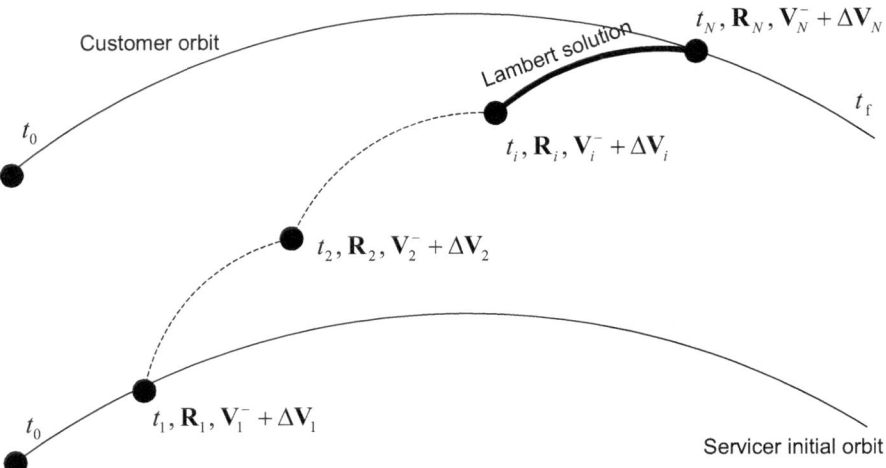

Fig. 3.9 Multi-impulse trajectory planning via random optimization

for direct children points is 2 and the value for children of these points is 3, and so on. *A*_ cost* is the sum of the total cost calculated to reach the waypoint; *out_degree* denotes the number of times the waypoint has been expanded.

The EXPAND_WAYPOINT method generates a new waypoint n which may be connected to the previous waypoint p. To satisfy motion constraints and to enable such connections with a simple discrete control, we apply a control to the waypoint p and integrate over a time. To explore the space, the control direction and magnitude as well as the time to integrate (impulse interval) are chosen randomly within a certain range. In this way, the cost can be easily calculated.

The ADD_TO_TREE method functions as follows. First, it checks whether the path between waypoints p and n is feasible. If not, abandon n. Second, if the path is feasible, the method calculates the cost to reach the new waypoint n (by summing the cost to get to the previous waypoint p and the cost from p to n). Third, the new waypoint n is then added to the tree as an outbranch from the previous waypoint p. Finally, the cost to reach the terminal state is computed, which is then used to check whether the path from the waypoint n to the terminal state is feasible.

In summary, the randomized A* tree expansion algorithm is capable of searching for a feasible solution quickly, on which basis the search can continue for an even better one. In this way, the search process can be flexibly controlled.

3.3.2 Planning Model

As depicted in Fig. 3.9, the last two impulses are computed by Lambert solution.

The specific steps of the randomized A* tree expansion algorithm are given as follows.

1. Design the weighting function for waypoints, written as

$$weight = \frac{order}{A^*\_cost \cdot out\_degree} \tag{3.22}$$

where $A^*\_\ cost = J$.
2. Generate randomly the mission duration ΔT, the time interval Δt from the current waypoint to the next one, and the required impulse velocity increment ΔV at the current point.
   Define ΔT_{min} as the required minimum mission duration, then

$$\Delta T = \Delta T_{min} + rand \cdot (t_f - t_0 - \Delta T_{min}) \tag{3.23}$$

where $rand$ is a uniformly distributed function in $(0,1)$.
   Define Δt_{min} as the minimum time interval between two adjacent impulses, and Δt_{pass} the duration from the initial state to the current waypoint, then

$$\Delta t = \Delta t_{min} + rand \cdot (\Delta T - \Delta t_{min} - \Delta t_{pass}) \tag{3.24}$$

Define ΔV_{min}, ΔV_{max} as the minimum and maximum magnitude of the impulse increment of the servicer, then the magnitude of ΔV is given as

$$\Delta V = \Delta V_{min} + rand \cdot (\Delta V_{max} - \Delta V_{min}) \tag{3.25}$$

Define θ_{max} as the feasible maximal angle between ΔV and the first impulse direction of the two-impulse maneuver from the waypoint to the terminal state, then the direction of ΔV is chosen by normal distribution in the range of $\pm \theta_{max}$.
3. Choose an orbit prediction model for the dynamics integration from the father waypoint to the children waypoint.
4. Check whether the generated children waypoint satisfies observation and control constraint. If so, this waypoint can be taken as a father and continues to generate more children waypoints; if not, we should go back to the father and regenerate a new one.
5. Solve the two-impulse maneuver problem from a waypoint to the terminal by Lambert solution.
6. Determine the best outbranch. We add all generated waypoints to the tree and examine every waypoint to see whether its last two impulses satisfy the impulse constraint. If so, this outbranch represents a feasible solution. Then among all feasible outbranches, we can find the best one based on the cost functions. On this best outbranch, the number of waypoints plus 1 is the optimized N. The ignition moment and velocity increment can be derived from the waypoint.

Table 3.3 Simulation results via random optimization

–	t (s)	ΔV (m/s)	ΔV_x (m/s)	ΔV_y (m/s)	ΔV_z (m/s)
Impulse-1	5,209	113.62	−40.78	−55.08	90.63
Impulse-2	14,352	112.27	12.65	35.00	−105.93
Impulse-3	30,972	69.22	60.95	31.44	−9.37
Impulse-4	47,892	24.02	−13.07	−20.11	1.26

3.3.3 Numerical Simulation

The simulation case is same as Sect. 3.2.3. Let $\Delta V_{min} = 10$ m/s, $\Delta t_{min} = 600$ s, $\theta_{max} = \pi/2$, and $Num = 900$. Each simulation yields different impulse times, ignition moments, and velocity increments, one of which is given as in Table 3.3.

The simulation shows that a feasible solution can be quickly found and the number of feasible solutions increases with the iteration times. Therefore, given sufficient computation time, the solution can be further optimized. Additionally, the convergence time can also be tuned by changing iteration times. Note that the resulting solution is still suboptimal.

References

1. Battin, R. H. (1999). *An introduction to the mathematics and methods of astrodynamics.* Reston, VA: AIAA.
2. Okuda, K., Yonemoto, K., & Akiyama, T. (2009). *Hybrid optimal trajectory generation using genetic algorithm and sequential quadratic programming.* ICROS-SICE International Joint Conference.
3. Ayala, H. V. H., & Coelho, L. S. (2008). A multiobjective genetic algorithm applied to multivariable control optimization. *ABCM Symposium Series in Mechatronics, 3,* 736–745.
4. Phillips, J. M., & Kavraki, L. E. (2003). *Spacecraft rendezvous and docking with real-time randomized optimization.* AIAA Guidance, Navigation, and Control Conference and Exhibit, Austin, Texas.
5. Saivea, G., & Vasileb, M. (2004). Probabilistic optimization applied to spacecraft rendezvous on Keplerian orbits. http://naca.central.cranfield.ac.uk/dcsss/2004. Accessed June 25, 2006.

Chapter 4
Proximity Relative Motion Planning

Abstract Proximity relative motion phase, whose planning is based on spacecraft relative motion dynamics, can best characterize on-orbit operations. First, the trajectory planning models and algorithms are examined based on three different thrust modes named impulse thrust, Bang-Bang thrust, and constant low thrust. Then, for a specific case of close proximity inspection, the problem of mission trajectory planning is addressed considering whether the customer is free. The mission constraints and 6-DOF coupled dynamic model of close proximity inspection are formulated. For a free customer, the hp-Adaptive Pseudospectral Method (hp-APM) is adopted to solve local inspection trajectory planning. Furthermore, if the customer structure is complicated, the Improved Artificial Potential Function (IAPF) is applied to surface following trajectory optimization when the field of view (FOV) is blocked. For a maneuvered customer, the Inverse Dynamics in the Virtual Domain method (IDVD) is explored to approach online trajectory planning.

4.1 Problem Formulation

Spacecraft relative motion trajectory planning is to determine an optimized transfer trajectory under various constraints based on the relative motion dynamic model, which is generally established by C-W equations or T-H equations.

The dynamic model is set up in the orbital reference frame of the customer, which is also called relative motion frame or Hill frame, defined with its origin attached to the customer center of mass, x axis aligned with the position vector and pointing toward the customer, y axis locating at orbital plane and perpendicular to x axis, and z axis completing the right-hand system.

As illustrated in Fig. 4.1, the servicer is enabled to achieve the desired state B from the initial state A in the mission duration Δt at the cost of ΔV. The objective of

L. Yang et al., *On-Orbit Operations Optimization: Modeling and Algorithms*,
SpringerBriefs in Optimization, DOI 10.1007/978-1-4939-0838-7_4,
© Leping Yang, Yanwei Zhu, Xianhai Ren, Yuanwen Zhang 2014

Fig. 4.1 Spacecraft
proximity relative motion

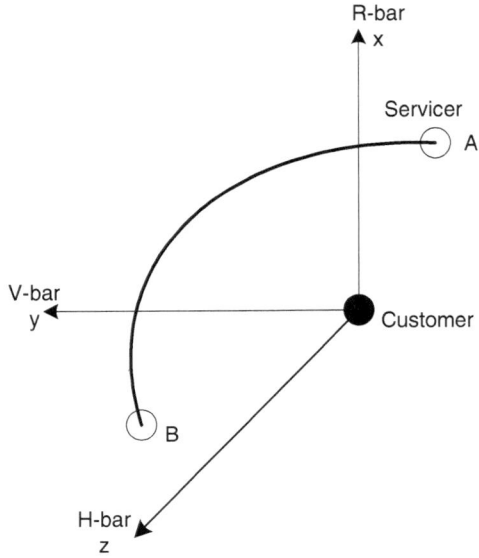

proximity relative motion planning is to optimize Δt and/or ΔV to generate an optimized relative motion trajectory.

The specific dynamic model is related to the customer orbit eccentricity. If the customer runs in a near-circular orbit, the relative motion dynamic model is given by C-W equations as

$$\begin{cases} \ddot{x} - 2n\dot{y} - 3n^2x = f_x \\ \ddot{y} + 2n\dot{x} = f_y \\ \ddot{z} + n^2z = f_z \end{cases} \tag{4.1}$$

where n denotes the customer mean orbital angular velocity, $\mathbf{f} = \mathbf{f}_d + \mathbf{F}_s/m_s$, $\mathbf{f}_d$ is relative perturbation acceleration, $\mathbf{F}_s$ the control force applied to the servicer, and m_s the servicer mass.

If the customer runs in an elliptical orbit, the first-order approximation expression of relative motion dynamic model is given as

$$\begin{cases} \ddot{x} - 2\dot{\theta}\dot{y} - \dot{\theta}^2x - \ddot{\theta}y - 2\mu x/r_c^3 = f_x \\ \ddot{y} + 2\dot{\theta}\dot{x} - \dot{\theta}^2y + \ddot{\theta}x + \mu y/r_c^3 = f_y \\ \ddot{z} + \mu z/r_c^3 = f_z \end{cases} \tag{4.2}$$

where r_c, e, θ are geocentric distance, orbital eccentricity, and true anomaly of the customer respectively, μ is Earth gravitational constant and $\mu/r_c^3 = n^2(1 + e \cos \theta)^3/(1 - e^2)^3$.

The solution to the above model derives the relationship between Δt and ΔV, whose specific formulation depends on the servicer thruster mode, which can be generally classified into impulse thrust, Bang-Bang thrust, and constant low thrust. Different thruster modes determine how do the cost function and control constraint specifically formulate.

In proximity relative motion missions, the spacecraft is subjected to various constraints, including state constraint and safety constraint. State constraint, which poses limitations on relative position or relative velocity, usually takes the form of line-of-sight (LOS) constraint and collision avoidance constraint, with the former being further divided into making the trajectory in or out of a certain region. The specific formulation of the state constraint is dependent on mission requirements. Safety constraint guarantees collision avoidance in the event of thruster failures, computer anomalies, and loss of sensing for higher robustness. In general, the shorter the relative distance is, the more complex the constraints are.

Therefore, the complexity of the planning model hinges on the eccentricity of customer orbit, relative distance, thruster mode, and mission constraint, leading to various forms of the planning model, which requires selecting a proper algorithm.

In the following sections, we attempt to address different forms of the planning model, which is divided by the relative distance, i.e., proximity and close proximity inspection. For the former, further classifications are made by thruster mode, elaborated in Sects. 4.2–4.4 respectively. For the latter, further classifications are made by mission constraints, elaborated in Sects. 4.5–4.7 respectively.

4.2 Sequential Quadratic Programming for Impulse Thrust Mode

Spacecraft relative motion trajectory planning with fixed N impulses is a typical constraint optimization problem, solvable by Sequential Quadratic Programming (SQP) algorithm. Impulse thrust hypothesis helps to simplify complicated problems. The following analysis is based on an example of two-impulse trajectory planning.

4.2.1 SQP Algorithm

To solve the constrained optimization problem, there are generally two kinds of approaches. One is to find the solution by the unconstrained optimization algorithm after previous disposal of constraints. In this approach, the key lies in choosing the penalty function and designing the penalty. However, it requires iteration and affords insufficient precision in equality constraint satisfaction. The other is to search for the optimal solution among all feasible ones, which is most typically represented by SQP. Previous applications show that SQP exhibits global convergence [1]. The basic principle of SQP is to transform the constrained optimization problem to a series of quadratic planning problems.

4.2.2 Two-Impulse Maneuver Model

The two-impulse maneuver model is derived by the dynamics equation in the form of state space.

4.2.2.1 Near-Circular Reference Orbit

Defining $\mathbf{r} = [\,x \quad y \quad z\,]^T$ and $\mathbf{v} = [\,\dot{x} \quad \dot{y} \quad \dot{z}\,]^T$ as relative position and velocity vectors, $\mathbf{x} = \begin{bmatrix} \mathbf{r}^T & \mathbf{v}^T \end{bmatrix}^T$, $\mathbf{u} = [\,a_x \quad a_y \quad a_z\,]^T$, and neglecting perturbation influences, C-W equations can be written as

$$\dot{\mathbf{x}}(t) = \mathbf{A}\mathbf{x}(t) + \mathbf{B}\mathbf{u}(t) \tag{4.3}$$

where

$$\mathbf{A} = \begin{bmatrix} 0 & 0 & 0 & 1 & 0 & 0 \\ 0 & 0 & 0 & 0 & 1 & 0 \\ 0 & 0 & 0 & 0 & 0 & 1 \\ 3n^2 & 0 & 0 & 0 & 2n & 0 \\ 0 & 0 & 0 & -2n & 0 & 0 \\ 0 & 0 & -n^2 & 0 & 0 & 0 \end{bmatrix}, \ \mathbf{B} = \begin{bmatrix} 0 & 0 & 0 & 1 & 0 & 0 \\ 0 & 0 & 0 & 0 & 1 & 0 \\ 0 & 0 & 0 & 0 & 0 & 1 \end{bmatrix}^T \tag{4.4}$$

Letting $\Delta t = t_2 - t_1$, $s = \sin(n\Delta t)$, and $c = \cos(n\Delta t)$, the state transition matrix of C-W equations is written as

$$\boldsymbol{\Phi}(t_2, t_1) = \boldsymbol{\Phi}(t_2 - t_1) = e^{\mathbf{A}\Delta t}$$

$$= \begin{bmatrix} 4 - 3c & 0 & 0 & s/n & 2(1-c)/n & 0 \\ 6(s - n\Delta t) & 1 & 0 & 2(c-1)/n & (4s - 3n\Delta t)/n & 0 \\ 0 & 0 & c & 0 & 0 & s/n \\ 3ns & 0 & 0 & c & 2s & 0 \\ 6n(c-1) & 0 & 0 & -2s & 4c - 3 & 0 \\ 0 & 0 & -ns & 0 & 0 & c \end{bmatrix}$$

$$= \begin{bmatrix} \boldsymbol{\Phi}_{rr} & \boldsymbol{\Phi}_{rv} \\ \boldsymbol{\Phi}_{vr} & \boldsymbol{\Phi}_{vv} \end{bmatrix} \tag{4.5}$$

which yields

$$\begin{bmatrix} \mathbf{r}_2 \\ \mathbf{v}_2 \end{bmatrix} = \begin{bmatrix} \boldsymbol{\Phi}_{rr} & \boldsymbol{\Phi}_{rv} \\ \boldsymbol{\Phi}_{vr} & \boldsymbol{\Phi}_{vv} \end{bmatrix} \begin{bmatrix} \mathbf{r}_1 \\ \mathbf{v}_1 \end{bmatrix} \tag{4.6}$$

Denote the superscripts "$-$" and "$+$" as the moment just before and after the ignition, then

$$\begin{cases} \mathbf{v}_1^+ = \boldsymbol{\Phi}_{rv}^{-1}(\mathbf{r}_2 - \boldsymbol{\Phi}_{rr}\mathbf{r}_1) \\ \mathbf{v}_2^- = \boldsymbol{\Phi}_{vr}\mathbf{r}_1 + \boldsymbol{\Phi}_{vv}\mathbf{v}_1^+ \end{cases} \tag{4.7}$$

where

$$\boldsymbol{\Phi}_{rv}^{-1}(\Delta t) = \begin{bmatrix} \dfrac{n(3\psi - 4\sin\psi)}{-8 + 8\cos\alpha + 3\alpha\sin\alpha} & \dfrac{-2n(\cos\psi - 1)}{-8 + 8\cos\psi + 3\psi\sin\psi} & 0 \\[4mm] \dfrac{2n(\cos\psi - 1)}{-8 + 8\cos\psi + 3\psi\sin\psi} & \dfrac{-n\sin\psi}{-8 + 8\cos\psi + 3\psi\sin\psi} & 0 \\[4mm] 0 & 0 & \dfrac{n}{\sin\psi} \end{bmatrix}, \ \psi = n\Delta t \tag{4.8}$$

Note that singular points would appear when ψ is an integer multiple of π, which should be avoided in computation [2].

Equation (4.7) derives the two-impulse vector as

$$\begin{cases} \Delta\mathbf{v}_1 = \mathbf{v}_1^+ - \mathbf{v}_1^- \\ \Delta\mathbf{v}_2 = \mathbf{v}_2^+ - \mathbf{v}_2^- \end{cases} \tag{4.9}$$

4.2.2.2   Elliptical Reference Orbit

Defining $k = \mu/h^{3/2}$ ($h = r^2\dot{\theta}$), Eq. (4.2) is transformed to [3]

$$\begin{cases} \ddot{x} - 2\dot{\theta}\dot{y} - \dot{\theta}^2 x - \ddot{\theta}y - 2k\omega^{3/2}x = 0 \\ \ddot{y} + 2\dot{\theta}\dot{x} - \dot{\theta}^2 y + \ddot{\theta}x + k\omega^{3/2}y = 0 \\ \ddot{z} + k\omega^{3/2}z = 0 \end{cases} \tag{4.10}$$

To derive the state transition matrix in a simple form, choose θ instead of t as the independent variable. Define $\rho = 1 + e\cos\theta$ and $\tilde{\mathbf{r}} = \rho\mathbf{r}$, then

$$\tilde{\mathbf{v}} = -e\sin\theta\mathbf{r} + \mathbf{v}/(k^2\rho) \tag{4.11}$$

Equation (4.10) can be rewritten as

$$\begin{cases} \tilde{x}'' = 2\tilde{z}' \\ \tilde{y}'' = -\tilde{y} \\ \tilde{z}'' = 3\tilde{z}/\rho - 2\tilde{x}' \end{cases} \tag{4.12}$$

Defining $\boldsymbol{\Phi}_{\theta_0}^{\theta}(t, t_0) = \boldsymbol{\Phi}_{\theta}\boldsymbol{\Phi}_{\theta_0}^{-1}$ as the state transition matrix of Eq. (4.12) from t_0 to t, its analytical solution can be expressed as

$$\begin{bmatrix} \tilde{\mathbf{r}}(\theta) \\ \tilde{\mathbf{v}}(\theta) \end{bmatrix} = \boldsymbol{\Phi}_{\theta}\boldsymbol{\Phi}_{\theta_0}^{-1} \begin{bmatrix} \tilde{\mathbf{r}}(\theta_0) \\ \tilde{\mathbf{v}}(\theta_0) \end{bmatrix} = \begin{bmatrix} \boldsymbol{\Phi}_{rr} & \boldsymbol{\Phi}_{rv} \\ \boldsymbol{\Phi}_{vr} & \boldsymbol{\Phi}_{vv} \end{bmatrix} \begin{bmatrix} \tilde{\mathbf{r}}(\theta_0) \\ \tilde{\mathbf{v}}(\theta_0) \end{bmatrix} \qquad (4.13)$$

$$\boldsymbol{\Phi}_{\theta} = \begin{bmatrix} 0 & s & 0 & c & 2-3esJ & 0 \\ 1 & c(1+1/\rho) & 0 & -s(1+1/\rho) & -3\rho^2 J & 0 \\ 0 & 0 & c/\rho & 0 & 0 & s/\rho \\ 0 & s' & 0 & c' & -3e(s'J+s/\rho^2) & 0 \\ 0 & -2s & 0 & e-2c & -3(1-2esJ) & 0 \\ 0 & 0 & -s/\rho & 0 & 0 & c/\rho \end{bmatrix} \quad (4.14)$$

$$\boldsymbol{\Phi}_{\theta_0}^{-1} = \frac{1}{1-e^2} \cdot$$

$$\begin{bmatrix} -3e\frac{s}{\rho}\left(1+\frac{1}{\rho}\right) & 1-e^2 & 0 & ec-2 & -es\left(1+\frac{1}{\rho}\right) & 0 \\ -3\frac{s}{\rho}\left(1+\frac{e^2}{\rho}\right) & 0 & 0 & c-2e & -s\left(1+\frac{1}{\rho}\right) & 0 \\ 0 & 0 & (1-e^2)\frac{c}{\rho} & 0 & 0 & -(1-e^2)\frac{s}{\rho} \\ -3\left(\frac{c}{\rho}+e\right) & 0 & 0 & -s & -c\left(1+\frac{1}{\rho}\right)-e & 0 \\ 3\rho+e^2-1 & 0 & 0 & es & \rho^2 & 0 \\ 0 & 0 & (1-e^2)\frac{s}{\rho} & 0 & 0 & (1-e^2)\frac{c}{\rho} \end{bmatrix}_{\theta_0}$$

$$(4.15)$$

where $s = \rho \sin \theta$, $c = \rho \cos \theta$, $s' = \cos \theta + e \cos 2\theta$, $c' = -\sin \theta - e \sin 2\theta$, θ_0, and θ are true anomaly of the customer at t_0 and t, and $J = k^2(t - t_0)$.

Recalling the transformed variables in Eq. (4.11), the inverse transformation is

$$\mathbf{r} = \tilde{\mathbf{r}}/\rho, \mathbf{v} = k^2(e \sin \theta \tilde{\mathbf{r}} + \rho \tilde{\mathbf{v}}) \qquad (4.16)$$

then

$$\begin{cases} \tilde{\mathbf{v}}_1^+ = \boldsymbol{\Phi}_{rv}^{-1}(\tilde{\mathbf{r}}_2 - \boldsymbol{\Phi}_{rr}\tilde{\mathbf{r}}_1) \\ \tilde{\mathbf{v}}_2^- = \boldsymbol{\Phi}_{vr}\tilde{\mathbf{r}}_1 + \boldsymbol{\Phi}_{vv}\tilde{\mathbf{v}}_1^+ \end{cases} \qquad (4.17)$$

By inverse transformation, the two-impulse vector is rewritten in t-domain as

$$\begin{cases} \Delta\mathbf{v}_1 = \mathbf{v}_1^+ - \mathbf{v}_1^- \\ \Delta\mathbf{v}_2 = \mathbf{v}_2^+ - \mathbf{v}_2^- \end{cases} \qquad (4.18)$$

4.2.3 Two-Impulse Trajectory Planning Model

Given the initial states t_0, $\mathbf{r}_0$, $\mathbf{v}_0$ and the terminal states $\mathbf{r}_f$, $\mathbf{v}_f$, two impulse vectors $\Delta \mathbf{v}_1$ and $\Delta \mathbf{v}_2$ are optimized with certain constraints.

1. Case 1: Fixed time mode
   In this mode, t_f is given and decision variables are t_1, t_2. The two-impulse planning model can be transformed to a constrained nonlinear optimization problem as

$$\min_{t_1, t_2} J = \Delta v_1 + \Delta v_2$$
$$\text{s.t. } t_0 \leq t_1 < t_2 \leq t_f$$
$$t_2 - t_1 \geq \Delta t_{min} \tag{4.19}$$
$$\Delta v_{min} \leq \Delta v_1 \leq \Delta v_{max}$$
$$\Delta v_{min} \leq \Delta v_2 \leq \Delta v_{max}$$

   where Δt_{min} denotes minimum impulse interval.
2. Case 2: Free time mode
   In this mode, the allowable maximum mission duration ΔT_{max} is given, and decision variables are t_1, t_2, and t_f. The planning model is

$$\min_{t_1, t_2, t_f} J = \eta(t_f - t_0)/\Gamma + (1 - \eta)(\Delta v_1 + \Delta v_2)$$
$$\text{s.t. } t_0 \leq t_1 < t_2 \leq t_f \leq t_0 + \Delta T_{max}$$
$$t_2 - t_1 \geq \Delta t_{min} \tag{4.20}$$
$$\Delta v_{min} \leq \Delta v_1 \leq \Delta v_{max}$$
$$\Delta v_{min} \leq \Delta v_2 \leq \Delta v_{max}$$

   where $\eta \in (0, 1)$ is weighting ratio and Γ is a user-defined constant which denotes the tolerable mission duration increase corresponding to 1 m/s velocity increment decrease. Δv_{min} and Δv_{max} are minimum and maximum velocity increment offered by each impulse.

   Note that Eqs. (4.19) and (4.20) are both typical SQP problems, which can be solved by the optimization function "fmincon" in MATLAB toolbox, but the initial value sensitiveness must be emphasized.

4.2.4 Numerical Simulation

We assume the customer runs in an orbit of $a = 6,828.140$ km, $e = 0.02$.

$$t_0 = 0, \mathbf{r}_0 = [-500 \quad -2,000 \quad 0]^T \text{m}, \mathbf{v}_0 = [-0.2 \quad 1.2 \quad 0]^T \text{m/s};$$
$$\mathbf{r}_f = [0 \quad -500 \quad 0]^T \text{m}, \mathbf{v}_f = [0.12 \quad 0 \quad 0]^T \text{m/s}.$$

Fig. 4.2 Optimized
two-impulse trajectory
for fixed time mode

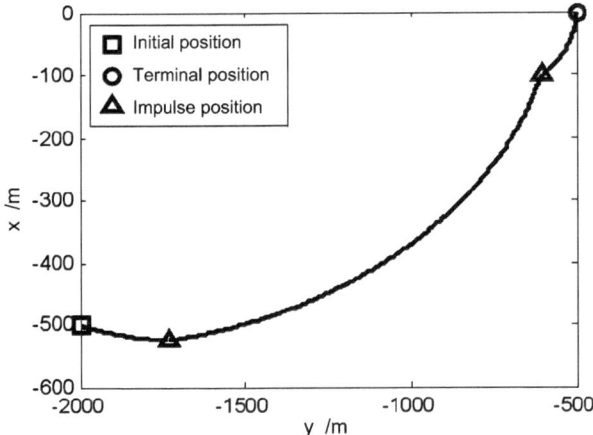

For a fixed time mission, $t_f = 3,600$ s; for a free time mission, $\Delta T_{max} = 3,600$ s, $\Delta v_{min} = 0.01$ m/s, $\Delta v_{max} = 0.2$ m/s, $\Gamma = 1,200$, $\eta = 0.2$, $\Delta t_{min} = 600$ s.

1. Case 1: Fixed time mode
   The optimized trajectory is presented as in Fig. 4.2, and the simulation results are given as follows.

$$t_1 = 216\,\text{s}, \mathbf{r}_1 = \begin{bmatrix} -524 & -1,734 & 0 \end{bmatrix}^T \text{m}, \Delta\mathbf{v}_1 = \begin{bmatrix} 0.1017 & -0.2220 & 0 \end{bmatrix}^T \text{m/s};$$

$$t_2 = 2,676\,\text{s}, \mathbf{r}_2 = \begin{bmatrix} -101 & -602 & 0 \end{bmatrix}^T \text{m}, \Delta\mathbf{v}_2 = \begin{bmatrix} 0.0892 & 0.1783 & 0 \end{bmatrix}^T \text{m/s}.$$

2. Case 2: Free time mode
   The optimized trajectory is presented as in Fig. 4.3, and the simulation results are given as follows.

$$t_1 = 226\,\text{s}, \mathbf{r}_1 = \begin{bmatrix} -525 & -1,722 & 0 \end{bmatrix}^T \text{m}, \Delta\mathbf{v}_1 = \begin{bmatrix} -0.0193 & -0.1596 & 0 \end{bmatrix}^T \text{m/s};$$

$$t_2 = 2,312\,\text{s}, \mathbf{r}_2 = \begin{bmatrix} 0 & -500 & 0 \end{bmatrix}^T \text{m}, \Delta\mathbf{v}_2 = \begin{bmatrix} -0.1467 & 0.1303 & 0 \end{bmatrix}^T \text{m/s};$$

$$t_f = 2,312\,\text{s}.$$

The simulation results have verified the validity of the proposed models and algorithm. It should be noted that it is the simplest case of spacecraft proximity relative motion problems, only suitable for preliminary mission design and analysis.

4.3 LP for Bang-Bang Thrust Mode

Considering engineering implementation, Bang-Bang thrust mode [4], i.e., fixed magnitude and adjustable actuation duration, is a common feasible assumption. Then the proximity relative motion trajectory planning could be addressed by Linear Programming (LP) based on a discrete dynamic model and linearized constraints.

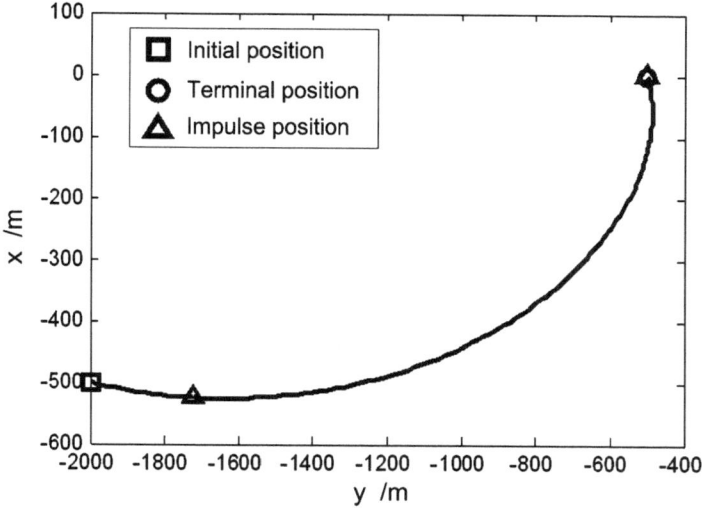

Fig. 4.3 Optimized two-impulse trajectory for free time mode

4.3.1 LP Algorithm

The standard LP problem can be formulated as

$$\min \mathbf{c}^{\mathrm{T}}\mathbf{x}$$
$$\text{s.t. } \mathbf{Ax} = \mathbf{b}, \mathbf{x} \geq \mathbf{0} \tag{4.21}$$

where

$$\mathbf{x} = \begin{bmatrix} x_1 \\ x_2 \\ \vdots \\ x_n \end{bmatrix}, \quad \mathbf{c} = \begin{bmatrix} c_1 \\ c_2 \\ \vdots \\ c_n \end{bmatrix}, \quad \mathbf{b} = \begin{bmatrix} b_1 \\ b_2 \\ \vdots \\ b_n \end{bmatrix}, \quad \mathbf{A} = \begin{bmatrix} a_{11} & a_{12} & \cdots & a_{1n} \\ a_{21} & a_{22} & \cdots & a_{2n} \\ \vdots & \vdots & & \vdots \\ a_{m1} & a_{m2} & \cdots & a_{mn} \end{bmatrix} \tag{4.22}$$

where $\mathbf{A}$ is the constraint matrix which satisfies $n \geq m \geq 1$.

However, the practical problem usually includes inequality constraint, nonconvex constraint, free decision variable, etc., which must be first approached to transform to a standard LP model.

4.3.1.1 Inequality Constraint

Introducing a slack variable x_{n+i} leads to the transformation of the inequality constraint

$$\sum_{j=1}^{n} a_{ij}x_j \leq b_i \tag{4.23}$$

to the equation constraint

$$\sum_{j=1}^{n} a_{ij}x_j + x_{n+i} = b_i, \quad x_{n+i} \geq 0 \tag{4.24}$$

4.3.1.2 Nonconvex Constraint

Introducing a large number M and a binary variable s_i leads to the transformation of the nonconvex constraint

$$\begin{aligned} &\sum_{j=1}^{n} a_{1j}x_j \leq b_1 \\ \text{OR } &\sum_{j=1}^{n} a_{2j}x_j \leq b_2 \\ \vdots\qquad &\qquad\vdots \\ \text{OR } &\sum_{j=1}^{n} a_{pj}x_j \leq b_p \end{aligned} \tag{4.25}$$

to the convex constraint

$$\begin{aligned} &\sum_{j=1}^{n} a_{1j}x_j \leq b_1 + Ms_1 \\ \text{AND } &\sum_{j=1}^{n} a_{2j}x_j \leq b_2 + Ms_2 \\ \vdots\qquad &\qquad\vdots \\ \text{AND } &\sum_{j=1}^{n} a_{pj}x_j \leq b_p + Ms_p \\ \text{AND } &\sum_{i=1}^{p} s_i \leq p - 1 \end{aligned} \tag{4.26}$$

4.3.1.3 Free Decision Variable

Introducing two additional variables x_k', x_k'' leads to the transformation of the free variable x_k to

$$x_k = x_k' - x_k'', \quad x_k' \geq 0, \quad x_k'' \geq 0 \tag{4.27}$$

If the planning model does not contain any nonconvex constraint, it can be solved by either the MATLAB linear programming function "linprog" or a combination of AMPL modeling language and ILOG CPLEX software [5]. If it contains nonconvex constraints, it is a Mixed-Integer Linear Programming (MILP) problem [6], which can be solved by using AMPL modeling language and ILOG CPLEX software.

4.3.2   Discrete Dynamic Model

The state space model of Eq. (4.2) is derived as

$$\dot{\mathbf{x}}(t) = \mathbf{A}_c(t)\mathbf{x}(t) + \mathbf{B}_c(t)\mathbf{u}(t) \tag{4.28}$$

where

$$\mathbf{A}_c = \begin{bmatrix} 0 & 0 & 0 & 1 & 0 & 0 \\ 0 & 0 & 0 & 0 & 1 & 0 \\ 0 & 0 & 0 & 0 & 0 & 1 \\ \dfrac{3 + e\cos\theta}{1 + e\cos\theta}\dot{\theta}^2 & \ddot{\theta} & 0 & 0 & 2\dot{\theta} & 0 \\ -\ddot{\theta} & \dfrac{e\cos\theta}{1 + e\cos\theta}\dot{\theta}^2 & 0 & -2\dot{\theta} & 0 & 0 \\ 0 & 0 & -\dfrac{1}{1 + e\cos\theta}\dot{\theta}^2 & 0 & 0 & 0 \end{bmatrix},$$

$$\mathbf{B}_c = \begin{bmatrix} 0 & 0 & 0 & 1 & 0 & 0 \\ 0 & 0 & 0 & 0 & 1 & 0 \\ 0 & 0 & 0 & 0 & 0 & 1 \end{bmatrix}^{\mathrm{T}}$$

$$\tag{4.29}$$

Applying zeroth-order holding mechanism to the system with a period of $T(t_0 = 0, t = kT)$, the discrete dynamic model is written as

$$\mathbf{x}(k+1) = \mathbf{A}_d(k)\mathbf{x}(k) + \mathbf{B}_d(k)\mathbf{u}(k) \tag{4.30}$$

where

$$\begin{aligned} \mathbf{A}_d(k) &= \mathbf{\Phi}((k+1)T, kT) \triangleq \mathbf{\Phi}(k+1, k) \\ \mathbf{B}_d(k) &= \int_{kT}^{(k+1)T} \mathbf{\Phi}((k+1)T, \tau)\mathbf{B}(\tau)\mathrm{d}\tau \end{aligned} \tag{4.31}$$

$$\mathbf{x}(k) = \mathbf{\Phi}(k,0)\mathbf{x}_0 + \sum_{j=0}^{k-1} \mathbf{\Phi}(k,j+1)\mathbf{B}_\mathrm{d}(j)\mathbf{u}(j) \tag{4.32}$$

where $\mathbf{\Phi}(k, m) = \mathbf{A}_\mathrm{d}(k - 1)\mathbf{A}_\mathrm{d}(k - 2) \ldots \mathbf{A}_\mathrm{d}(m)$.

For the circular reference orbit, Eq. (4.30) denotes a linear time-invariant system, then

$$\mathbf{A}_\mathrm{d} = \begin{bmatrix} 4 - 3\cos nT & 0 & 0 & \dfrac{\sin nT}{n} & \dfrac{2(1 - \cos nT)}{n} & 0 \\ 6(\sin nT - nT) & 1 & 0 & \dfrac{2(\cos nT - 1)}{n} & \dfrac{(4\sin nT - 3nT)}{n} & 0 \\ 0 & 0 & \cos nT & 0 & 0 & \dfrac{\sin nT}{n} \\ 3n\sin nT & 0 & 0 & \cos nT & 2\sin nT & 0 \\ 6n(\cos nT - 1) & 0 & 0 & -2\sin nT & 4\cos nT - 3 & 0 \\ 0 & 0 & -n\sin nT & 0 & 0 & \cos nT \end{bmatrix} \tag{4.33}$$

$$\mathbf{B}_\mathrm{d} = \begin{bmatrix} (1 - \cos nT)/n^2 & 2(nT - \sin nT)/n^2 & 0 \\ 2(\sin nT - nT)/n^2 & 4(1 - \cos nT)/n^2 - 3T^2/2 & 0 \\ 0 & 0 & (1 - \cos nT)/n^2 \\ \sin nT/n & 2(1 - \cos nT)/n & 0 \\ 2(\cos nT - 1)/n & (4\sin nT - 3nT)/n & 0 \\ 0 & 0 & \sin nT/n \end{bmatrix} \tag{4.34}$$

$$\mathbf{x}(k) = \mathbf{A}_\mathrm{d}^k \mathbf{x}_0 + \begin{bmatrix} \mathbf{A}_\mathrm{d}^{k-1}\mathbf{B}_\mathrm{d} & \mathbf{A}_\mathrm{d}^{k-2}\mathbf{B}_\mathrm{d} & \cdots & \mathbf{A}_\mathrm{d}\mathbf{B}_\mathrm{d} & \mathbf{B}_\mathrm{d} \end{bmatrix} \begin{bmatrix} \mathbf{u}(0) \\ \vdots \\ \mathbf{u}(k-1) \end{bmatrix} \tag{4.35}$$

Define $\mathbf{G}(k) = \begin{bmatrix} \mathbf{A}_\mathrm{d}^{k-1}\mathbf{B}_\mathrm{d} & \mathbf{A}_\mathrm{d}^{k-2}\mathbf{B}_\mathrm{d} & \cdots & \mathbf{A}_\mathrm{d}\mathbf{B}_\mathrm{d} & \mathbf{B}_\mathrm{d} \end{bmatrix}$ and $\mathbf{U}(k) = \begin{bmatrix} \mathbf{u}^\mathrm{T}(0) & \cdots \\ \end{bmatrix}$ $\mathbf{u}^\mathrm{T}(k-1)]^\mathrm{T}$, then

$$\mathbf{x}(k) = \mathbf{A}_\mathrm{d}^k \mathbf{x}_0 + \mathbf{G}(k)\mathbf{U}(k) \tag{4.36}$$

For the elliptical reference orbit, Eq. (4.28) denotes a linear time-varying system, which is difficult to discretize and usually takes a first-order approximation.

$$\mathbf{\Phi}(k+1,k) \approx \mathbf{I} + \mathbf{A}_\mathrm{c}(kT) \cdot T, \quad \mathbf{B}_\mathrm{d}(k) \approx \mathbf{B}_\mathrm{c}(kT) \cdot T \tag{4.37}$$

4.3.3 Constraint Linearization

Here we only consider three kinds of constraints, i.e., state constraint, control constraint, and safety constraint, which are linearized respectively as follows.

4.3.3.1 State Constraint

1. LOS constraint

   To make sure the relative trajectory is always within the inspection field of the customer, it is required that the trajectory at each time step be within a certain convex region, which is called LOS constraint. To guarantee the trajectories between discrete points in LOS, the FOV needs to be shrunk to some extent.

   Using polyhedron approximation, which is composed of N_L half-plane spaces [5], the LOS constraint can be expressed as

   $$\forall k = 1, \ldots, N_F : \quad \begin{matrix} a_1 x(k) + b_1 y(k) + c_1 z(k) \le d_1 \\ \text{AND} \quad a_2 x(k) + b_2 y(k) + c_2 z(k) \le d_2 \\ \vdots \qquad\qquad\qquad \vdots \\ \text{AND} \quad a_{N_L} x(k) + b_{N_L} y(k) + c_{N_L} z(k) \le d_{N_L} \end{matrix} \qquad (4.38)$$

   Define

   $$\mathbf{A}_L = \begin{bmatrix} a_1 & b_1 & c_1 \\ \vdots & \vdots & \vdots \\ a_{N_L} & b_{N_L} & c_{N_L} \end{bmatrix}, \quad \mathbf{d}_L = \begin{bmatrix} d_1 \\ \vdots \\ d_{N_L} \end{bmatrix}, \quad \mathbf{C} = \begin{bmatrix} 1 & 0 & 0 & 0 & 0 & 0 \\ 0 & 1 & 0 & 0 & 0 & 0 \\ 0 & 0 & 1 & 0 & 0 & 0 \end{bmatrix} \qquad (4.39)$$

   then

   $$\mathbf{A}_L \mathbf{C} \mathbf{x}(k) \le \mathbf{d}_L \qquad (4.40)$$

   which is substituted into Eq. (4.36) to give the function of $\mathbf{U}(k)$

   $$\mathbf{A}_L \mathbf{C} \mathbf{G}(k) \mathbf{U}(k) \le \mathbf{d}_L - \mathbf{A}_L \mathbf{C} \mathbf{A}_d^k \mathbf{x}_0 \qquad (4.41)$$

   which leads to

   $$\overline{\mathbf{S}}_L(k)_{N_L \times 3N_F} (\mathbf{U}_{N_F})_{3N_F \times 1} \le \overline{\mathbf{T}}_L(k)_{N_L \times 1} \quad k = 1, \ldots, N_F \qquad (4.42)$$

   where

   $$\mathbf{U}_{N_F} = \mathbf{U}(N_F) = \begin{bmatrix} \mathbf{u}^T(0) & \cdots & \mathbf{u}^T(N_F - 1) \end{bmatrix}^T \qquad (4.43)$$

2. Collision avoidance constraint

The concept of no-fly zone is introduced to avoid spacecraft collision. No matter what does the no-fly zone look like, it can be linearized by polyhedron approximation. For a convex polyhedron, the collision avoidance constraint can be expressed as [7]

$$
\forall k = 1, \ldots, N_F : \quad
\begin{aligned}
& a_{o1}x(k) + b_{o1}y(k) + c_{o1}z(k) \le d_{o1} \\
\text{OR} \quad & a_{o2}x(k) + b_{o2}y(k) + c_{o2}z(k) \le d_{o2} \\
& \vdots \qquad\qquad\qquad\qquad \vdots \\
\text{OR} \quad & a_{oN_O}x(k) + b_{oN_O}y(k) + c_{oN_O}z(k) \le d_{oN_O}
\end{aligned}
\tag{4.44}
$$

Introducing a binary variable $s_{oi}(k)$ ($i = 1, \ldots, N_O$) and a large number M leads to the transformation of "OR" constraint to "AND" constraint, written as

$$
\begin{aligned}
& a_{o1}x(k) + b_{o1}y(k) + c_{o1}z(k) \le d_{o1} + Ms_{o1}(k) \\
\text{AND} \quad & a_{o2}x(k) + b_{o2}y(k) + c_{o2}z(k) \le d_{o2} + Ms_{o2}(k) \\
& \vdots \qquad\qquad\qquad\qquad \vdots \\
\text{AND} \quad & a_{oN_O}x(k) + b_{oN_O}y(k) + c_{oN_O}z(k) \le d_{oN_O} + Ms_{oN_O}(k) \\
\text{AND} \quad & \sum_{i=1}^{N_O} s_{oi}(k) \le N_O - 1
\end{aligned}
\tag{4.45}
$$

Define

$$
\mathbf{A_O} = \begin{bmatrix} a_{o1} & b_{o1} & c_{o1} \\ \vdots & \vdots & \vdots \\ a_{oN_O} & b_{oN_O} & c_{oN_O} \end{bmatrix}, \quad
\mathbf{d_O} = \begin{bmatrix} d_{o1} \\ \vdots \\ d_{oN_O} \end{bmatrix}, \quad
\mathbf{s_O} = \begin{bmatrix} s_{o1} \\ \vdots \\ s_{oN_O} \end{bmatrix}
\tag{4.46}
$$

then

$$
\begin{cases}
\mathbf{A_O}\mathbf{C}\mathbf{x}(k) \le \mathbf{d_O} + M\mathbf{s_O}(k) \\
\|\mathbf{s_O}(k)\|_1 \le N_O - 1
\end{cases}
\tag{4.47}
$$

which is substituted into Eq. (4.36) to give

$$
\begin{cases}
\mathbf{A_O}\mathbf{C}\mathbf{G}(k)\mathbf{U}(k) \le \mathbf{d_O} - \mathbf{A_O}\mathbf{C}\mathbf{A}_d^k\mathbf{x_0} + M\mathbf{s_O}(k) \\
\|\mathbf{s_O}(k)\|_1 \le N_O - 1
\end{cases}
\tag{4.48}
$$

which leads to

$$\begin{cases} \overline{\mathbf{S}}_O(k)_{N_O \times 3N_F} (\mathbf{U}_{N_F})_{3N_F \times 1} \leq \overline{\mathbf{T}}_O(k)_{N_O \times 1} + M\mathbf{s}_O(k) \\ \|\mathbf{s}_O(k)\|_1 \leq N_O - 1 \end{cases} \quad k = 1, \ldots, N_F \quad (4.49)$$

For a nonconvex polyhedron, the above procedure can also apply with the only difference being in the addition of "AND" constraints. To prevent the trajectories between discrete points from intersecting with the no-fly zone, the concept of safety envelope is introduced so that the no-fly zone can be reasonably enlarged. In the x–y plane, the no-fly zone is generally enlarged by $(\sqrt{2}/2)v_{max}T$, while along the z direction by $\max(\dot{z}_{max}, |\dot{z}_{min}|)T$.

To improve computation efficiency, constraints can be strengthened to transform the nonconvex constraint of collision avoidance to a convex one (which means to keep the servicer in a certain customer-free region).

4.3.3.2 Control Constraint

The control constraint characterizes the thrust actuation in every direction [5]. If the thrust lasts for a time step T, the corresponding maximum thrust is u_{max}. If the thrust lasts for only an impulse bit, the corresponding minimum thrust is u_{min}.

1. Control input saturation constraint (convex constraint)
   The control input saturation constraint is imposed to ensure control inputs $\mathbf{u}$ to be within the capability of the actuators $\forall k = 1, \ldots, N_F$:

$$\mathbf{u}_L \leq \mathbf{u}(k) \leq \mathbf{u}_R \quad (4.50)$$

which leads to

$$\begin{bmatrix} \mathbf{I}_{3N_F \times 3N_F} \\ -\mathbf{I}_{3N_F \times 3N_F} \end{bmatrix} (\mathbf{U}_{N_F})_{3N_F \times 1} \leq \begin{bmatrix} \mathbf{U}_R \\ -\mathbf{U}_L \end{bmatrix} \quad (4.51)$$

where $\mathbf{I}$ denotes unit matrix, $\mathbf{u}_R$ and $\mathbf{u}_L$ refer to the upper and lower boundary, which usually satisfy $\mathbf{u}_L = -\mathbf{u}_R$. This constraint represents the maximum inputs of the thruster, written as

$$|\mathbf{u}(k)| = \mathbf{v}_k \leq \mathbf{u}_{max} = [u_{max\,x} \quad u_{max\,y} \quad u_{max\,z}]^T = \mathbf{u}_R \quad (4.52)$$

2. Minimum impulse bit constraint (nonconvex constraint)
   This constraint represents the minimum inputs of the thruster, written as

$$|\mathbf{u}(k)| = \mathbf{v}_k \geq \mathbf{u}_{min} = [u_{min\,x} \quad u_{min\,y} \quad u_{min\,z}]^T \quad (4.53)$$

Introducing a slack variable v_{kj} and a binary variable d_{kj} gives

$$d_{kj} u_{\min j} \leq v_{kj} \leq d_{kj} u_{\max j} \qquad (4.54)$$

when $d_{kj} = 0$, $v_{kj} = 0$, $u_{kj} = 0$; otherwise, $u_{\min j} \leq v_{kj} \leq u_{\max j}$.
    Denote $\mathbf{d}_k = \operatorname{diag}(d_{kx} \quad d_{ky} \quad d_{kz})$, then

$$\mathbf{d}_k \mathbf{u}_{\min} \leq \mathbf{v}_k \leq \mathbf{d}_k \mathbf{u}_{\max} \qquad (4.55)$$

4.3.3.3   Safety Constraint

When the servicer encounters any malfunction at the N_P step, it soon turns to the passive safety mode and shuts off all thrusters to prevent the forthcoming N_S steps from potential collision [8]. Generally, N_P should be close to N_F, which means the majority of control commands have already been executed, hence less restrictive on trajectory optimization. Otherwise, it imposes stronger restrictions on trajectory optimization, leading to a conservative result and low fuel efficiency.

Consider the safety constraint when failure occurs at the N_P step, then

$$\begin{aligned}
\mathbf{u}(N_P) &= \mathbf{u}(N_P + 1) = \cdots = \mathbf{u}(N_F - 1) = \mathbf{0} \\
\mathbf{x}_{\mathrm{FP}}(N_P) &= \mathbf{A}_{\mathrm{d}}^{N_P} \mathbf{x}_0 + \mathbf{G}(N_P) \mathbf{U}(N_P) \\
\forall k &> N_P, \mathbf{x}_{\mathrm{FP}}(k) = \mathbf{A}_{\mathrm{d}}^{k - N_P} \mathbf{x}_{\mathrm{FP}}(N_P)
\end{aligned} \qquad (4.56)$$

which transform the safety constraint to

$$\forall k \in \{N_P + 1, \ldots, N_P + N_S\} : \mathbf{x}_{\mathrm{FP}}(k) \notin \mathcal{T} \qquad (4.57)$$

where $\mathcal{T}$ implies the no-fly zone.

Similar to the previous collision avoidance constraint, the safety constraint can also be written as

$$\forall k = N_P + 1, \ldots, N_P + N_S : \begin{cases} \overline{\mathbf{S}}_{\mathrm{OP}}(k)_{N_O \times 3N_F} (\mathbf{U}_{N_F})_{3N_F \times 1} \leq \overline{\mathbf{T}}_{\mathrm{OP}}(k)_{N_O \times 1} + M \mathbf{s}_{\mathrm{OP}}(k) \\ \|\mathbf{s}_{\mathrm{OP}}(k)\|_1 \leq N_O - 1 \end{cases}$$
$$(4.58)$$

4.3.4   Planning Model

The cost function is written as

$$J = \sum_{k=0}^{N_F - 1} \|\mathbf{u}(k)\|_1 = \|\mathbf{U}_{N_F}\|_1 \qquad (4.59)$$

To find the optimized solution by LP or MILP, two nonnegative slack variables $\mathbf{U}_{N_F}^+$ and $\mathbf{U}_{N_F}^-$ are introduced to define the positive and negative parts of the control input

$$\mathbf{U}_{N_F} = \mathbf{U}_{N_F}^+ - \mathbf{U}_{N_F}^-, \quad \mathbf{U}_{N_F}^+ \geq \mathbf{0}, \quad \mathbf{U}_{N_F}^- \geq \mathbf{0} \tag{4.60}$$

Then, the control input vector is

$$\left(\hat{\mathbf{U}}_{N_F}\right)_{6N_F \times 1} = \left[\left(\mathbf{U}_{N_F}^+\right)^{\mathrm{T}} \quad \left(\mathbf{U}_{N_F}^-\right)^{\mathrm{T}} \right]^{\mathrm{T}} \tag{4.61}$$

The minimum impulse bit constraint given by Eq. (4.55) can be expressed as

$$\hat{\mathbf{D}} \hat{\mathbf{U}}_{\min} \leq \hat{\mathbf{U}}_{N_F} \leq \hat{\mathbf{D}} \hat{\mathbf{U}}_{\max} \tag{4.62}$$

where

$$\begin{aligned}
\hat{\mathbf{D}} &= \operatorname{diag}(\mathbf{d}_1 \quad \cdots \quad \mathbf{d}_{N_F} \quad \mathbf{d}_1 \quad \cdots \quad \mathbf{d}_{N_F}) \\
\left(\hat{\mathbf{U}}_{\min}\right)_{6N_F} &= \left[\mathbf{u}_{\min}^{\mathrm{T}} \quad \cdots \quad \mathbf{u}_{\min}^{\mathrm{T}}\right]^{\mathrm{T}}, \hat{\mathbf{U}}_{\max} = \left[\mathbf{U}_{\mathrm{R}}^{\mathrm{T}} \quad \mathbf{U}_{\mathrm{R}}^{\mathrm{T}}\right]^{\mathrm{T}}
\end{aligned} \tag{4.63}$$

In summary, the planning model can be written in the standard LP or MILP form as

$$\begin{aligned}
J^* &= \min_{\hat{\mathbf{U}}_{N_F}} \boldsymbol{\gamma}^{\mathrm{T}} \hat{\mathbf{U}}_{N_F} \\
\text{s.t.} \quad & \left[\mathbf{G}(N_F) \quad -\mathbf{G}(N_F)\right] \hat{\mathbf{U}}_{N_F} = \mathbf{x}_f - \mathbf{A}_d^{N_F} \mathbf{x}_0 \\
& \begin{bmatrix} -\mathbf{I}_{3N_F \times 3N_F} & \mathbf{0} \\ \mathbf{0} & -\mathbf{I}_{3N_F \times 3N_F} \end{bmatrix} \hat{\mathbf{U}}_{N_F} \leq \begin{bmatrix} \mathbf{0} \\ \mathbf{0} \end{bmatrix} \\
& \boldsymbol{\Lambda} \hat{\mathbf{U}}_{N_F} \leq \boldsymbol{\beta}
\end{aligned} \tag{4.64}$$

where $\boldsymbol{\gamma}$ denotes $6N_F \times 1$ unit vector, and $\boldsymbol{\Lambda} \hat{\mathbf{U}}_{N_F} \leq \boldsymbol{\beta}$ represents constraints mentioned in Sect. 4.3.3.

4.3.5   Numerical Simulation

Considering the independence between in-plane and out-of-plane motions, we only take into account the former case. Assume the customer runs in a circular orbit of 450 km.

$$\begin{aligned}
& u_{\max\,x} = 1.0 \times 10^{-3} \text{ m/s}^2, u_{\max\,y} = 2.0 \times 10^{-3} \text{ m/s}^2, T = 60 \text{ s}; \\
& t_0 = 0 \text{ s}, \mathbf{r}_0 = [-10 \quad 120]^{\mathrm{T}} m, \mathbf{v}_0 = [0.24 \quad 0.06]^{\mathrm{T}} m/s; \\
& t_f = 1,200 \text{ s}, \mathbf{r}_f = [0 \quad 10]^{\mathrm{T}} m, \mathbf{v}_f = [0 \quad 0]^{\mathrm{T}} m/s; \\
& \mathbf{A}_{\mathrm{L}} = \begin{bmatrix} 1 & -\tan(\pi/6) \\ -1 & -\tan(\pi/6) \end{bmatrix}, \mathbf{d}_{\mathrm{L}} = \begin{bmatrix} -5\tan(\pi/6) \\ -5\tan(\pi/6) \end{bmatrix}.
\end{aligned}$$

Fig. 4.4 In-plane
optimized trajectory
without considering safety
constraint

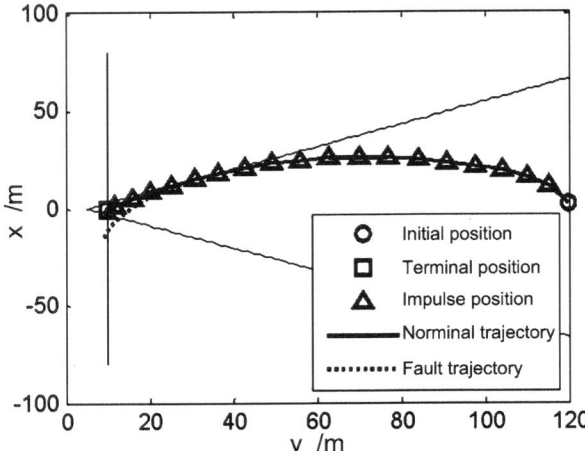

Fig. 4.5 Control
acceleration without
considering safety
constraint

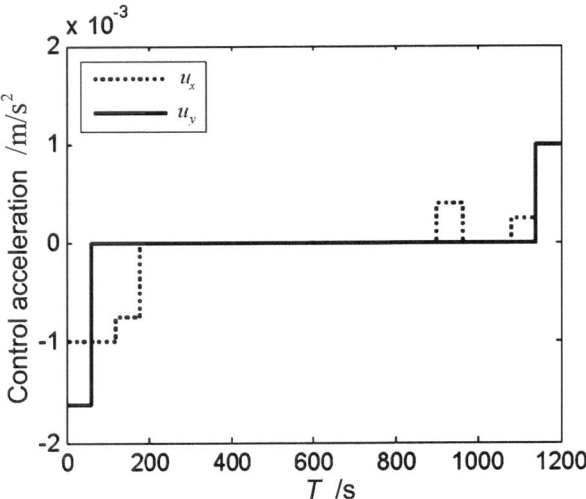

If we only consider the constraints of LOS and control input saturation, the
in-plane optimized transfer trajectory is generated as Fig. 4.4, the corresponding
control acceleration as Fig. 4.5, and fuel consumption is 0.420 m/s. The dotted line
of Fig. 4.5 shows the trajectory after a failure at $N_P = 15$.

If the safety constraint is added (limit the trajectory after the failure at $N_P = 15$
by the side of $y \geq 10$), given $N_S = 10$, the in-plane optimized transfer trajectory is
generated as Fig. 4.6, the corresponding control acceleration as Fig. 4.7, and fuel
consumption is 0.426 m/s, which indicates an increase as compared with the
previous case.

Fig. 4.6 In-plane optimized trajectory considering safety constraint

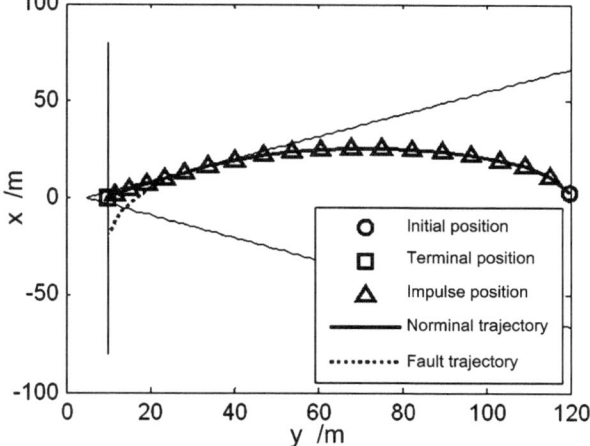

Fig. 4.7 Control acceleration considering safety constraint

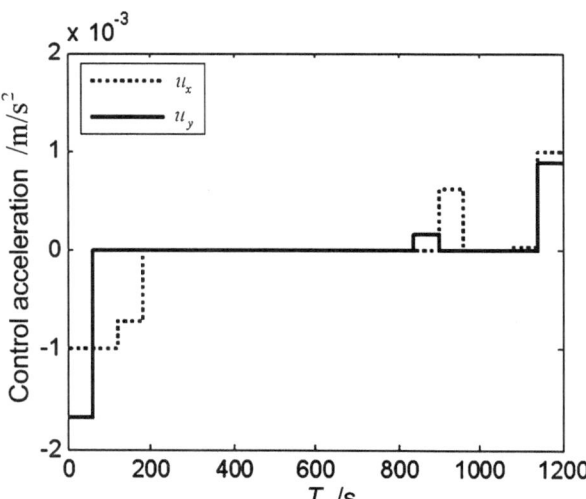

4.4   Pontryagin's Maximum Principle for Constant Low Thrust Mode

With the development of electric propulsion, continuous constant low thrust mode, i.e., small magnitude but high impulse bit, has become common, which can be addressed by the Pontryagin's Maximum Principle. The analysis in this section is based on a circular reference orbit assumption.

4.4.1 Pontryagin's Maximum Principle

The state model of the controlled system is written as

$$\dot{\mathbf{x}} = \mathbf{f}(\mathbf{x}, \mathbf{u}, t), \mathbf{x} \in \mathbb{R}^n, \mathbf{u} \in \mathbb{R}^m, m \le n \tag{4.65}$$

where $\mathbf{u}$ is defined in the bounded closed set $\mathcal{U}$. The terminal constraint is $\mathbf{G}[\mathbf{x}(t_f), t_f] = \mathbf{0}$ and the cost function is

$$\min J = \phi[\mathbf{x}(t_f), t_f] + \int_{t_0}^{t_f} F(\mathbf{x}, \mathbf{u}, t)\mathrm{d}t \tag{4.66}$$

Define the Hamiltonian as

$$H(\mathbf{x}, \mathbf{u}, \lambda, t) = F(\mathbf{x}, \mathbf{u}, t) + \lambda^{\mathrm{T}}\mathbf{f}(\mathbf{x}, \mathbf{u}, t) \tag{4.67}$$

then the necessary condition for $\min J$ is that $\mathbf{x}(t)$, $\mathbf{u}(t)$, $\lambda(t)$, and t_f satisfy the following equations.

1. Regular equations

$$\dot{\lambda} = -\frac{\partial H}{\partial \mathbf{x}}, \quad \dot{\mathbf{x}} = \frac{\partial H}{\partial \lambda} \tag{4.68}$$

   where λ is the costate vector.
2. Minimum conditions

$$H(\mathbf{x}^*, \mathbf{u}^*, \lambda, t) = \min_{\mathbf{u} \in \mathcal{U}} H(\mathbf{x}^*, \mathbf{u}, \lambda, t) \tag{4.69}$$

   where $\mathbf{x}^*$ is the optimal trajectory and $\mathbf{u}^*$ optimal control.
3. Boundary conditions

$$\mathbf{x}(t_0) = \mathbf{x}_0, \quad \mathbf{G}[\mathbf{x}(t_f), t_f] = \mathbf{0} \tag{4.70}$$

4. Transversal conditions

$$\lambda(t_f) = \frac{\partial \phi}{\partial \mathbf{x}(t_f)} + \frac{\partial \mathbf{G}^{\mathrm{T}}}{\partial \mathbf{x}(t_f)} \upsilon \tag{4.71}$$

   where υ is the column vector to be determined.
5. Hamiltonian
   For a time-invariant system, it follows

$$\begin{cases} t_f \text{ fixed} & H \equiv \text{constant} \\ t_f \text{ free} & H \equiv 0 \end{cases} \tag{4.72}$$

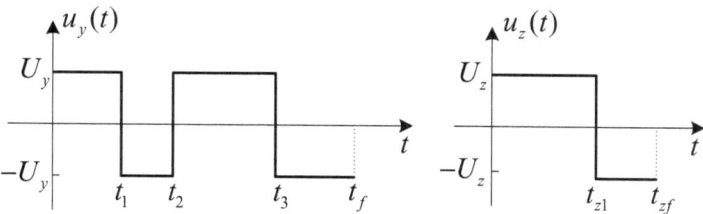

Fig. 4.8 Minimum-time maneuver control curve

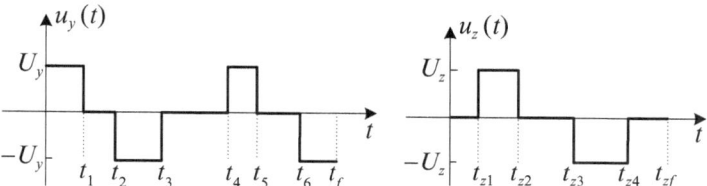

Fig. 4.9 Minimum-fuel maneuver control curve

For a time-varying system, the Hamiltonian is not constant and satisfies $H(t_f) = 0$.

According to Pontryagin's Maximum Principle, the minimum-time solution is Bang-Bang, whereas the minimum-fuel solution can be described by a relay function with a dead zone. It is well known that, for stable n-state systems, the number of required switch times is $n - 1$. Since the in-plane relative motion is controllable when only a transversal force is actuated, we assume that only transversal and normal forces act on the servicer to study minimum-fuel maneuver and minimum-time maneuver [9].

The minimum-time maneuver problem is a variable time state transfer problem, whose control curve is given as in Fig. 4.8, showing that the control applied at t_0 has two possible values $\pm U$. The left one corresponds to in-plane motion with $t_1 \sim t_3$ and t_f to be determined; the right one to out-of-plane motion with t_{z1} and t_{zf} to be determined. The minimum maneuver time is $\Delta t_{min} = \max\{t_f, t_{zf}\}$.

The minimum-fuel maneuver problem is a fixed time state transfer problem, whose control curve is given as in Fig. 4.9. The left one corresponds to in-plane motion with a given t_f and $t_1 \sim t_6$ to be determined; the right one to out-of-plane motion with a given t_{zf} and $t_{z1} \sim t_{z4}$ to be determined. Generally, $t_f = t_{zf} > t_0 + \Delta t_{min}$.

4.4.2 Dynamic Model

With a constant thrust $\mathbf{u} = \begin{bmatrix} u_x & u_y & u_z \end{bmatrix}^T$, the analytical solution to the C-W equations is written as

$$\mathbf{x}(t) = \mathbf{\Phi}(t_0, t)\mathbf{x}_0 + \mathbf{\Gamma}(t_0, t)\mathbf{u} \tag{4.73}$$

where $\mathbf{\Gamma}(t_0, t)$ is the control response matrix from t_0 to t. Define $\Delta t = t - t_0$, $s = \sin(n\Delta t)$, and $c = \cos(n\Delta t)$, then

$$\mathbf{\Gamma}(t_0, t) = \begin{bmatrix} (1-c)/n^2 & 2(n\Delta t - s)/n^2 & 0 \\ 2(s - n\Delta t)/n^2 & 4(1-c)/n^2 - 3\Delta t^2/2 & 0 \\ 0 & 0 & (1-c)/n^2 \\ s/n & 2(1-c)/n & 0 \\ 2(c-1)/n & (4s - 3n\Delta t)/n & 0 \\ 0 & 0 & s/n \end{bmatrix} \tag{4.74}$$

4.4.3 Planning Model

Here two cases of minimum-time maneuver and minimum-fuel maneuver are discussed, formulating in-plane and out-of-plane motions respectively for each case.

4.4.3.1 Minimum-Time Maneuver

The cost function is

$$\min J_{\text{time}} = t_{\text{f}} - t_0 \tag{4.75}$$

The Hamiltonian is

$$H_{\text{time}}(t) = 1 + \boldsymbol{\lambda}^{\mathrm{T}}(t)[\mathbf{A}\mathbf{x}(t) + \mathbf{B}\mathbf{u}(t)] \tag{4.76}$$

The costate equation is

$$\dot{\boldsymbol{\lambda}}(t) = -\mathbf{A}^{\mathrm{T}}\boldsymbol{\lambda}(t) \tag{4.77}$$

The minimum condition is

$$\boldsymbol{\lambda}^{\mathrm{T}}(t_i)\mathbf{B} = 0, \quad \forall t_i \tag{4.78}$$

where t_i is the switch time and $\boldsymbol{\lambda}^{\mathrm{T}}(t)\mathbf{B}$ switch function.
The Hamiltonian satisfies

$$H_{\text{time}}(t) = 0, \forall t \tag{4.79}$$

1. In-plane motion
   If only a transversal control is applied, $\mathbf{u}(t) = \begin{bmatrix} 0 & u_y(t) & 0 \end{bmatrix}^{\mathrm{T}}$, then

$$
\mathbf{x} = \begin{bmatrix} x \\ y \\ \dot{x} \\ \dot{y} \end{bmatrix}, \quad \mathbf{A} = \begin{bmatrix} 0 & 0 & 1 & 0 \\ 0 & 0 & 0 & 1 \\ 3n^2 & 0 & 0 & 2n \\ 0 & 0 & -2n & 0 \end{bmatrix}, \quad \mathbf{B} = \begin{bmatrix} 0 \\ 0 \\ 0 \\ 1 \end{bmatrix} \tag{4.80}
$$

Bang-Bang control assumes the control curve given as the left one in Fig. 4.8, in which the initial control is $u_{y0} = \pm U_y$. Equation (4.73) can then yield

$$
\begin{bmatrix} 4 - 3\cos T_f & 0 & -\sin T_f/n & 2(1 - \cos T_f)/n \\ 6(T_f - \sin T_f) & 1 & 2(\cos T_f - 1)/n & (3T_f - 4\sin T_f)/n \\ -3n\sin T_f & 0 & \cos T_f & -2\sin T_f \\ 6n(\cos T_f - 1) & 0 & 2\sin T_f & 4\cos T_f - 3 \end{bmatrix} \begin{bmatrix} x_f \\ y_f \\ \dot{x}_f \\ \dot{y}_f \end{bmatrix}
$$

$$
= \begin{bmatrix} x_0 \\ y_0 \\ \dot{x}_0 \\ \dot{y}_0 \end{bmatrix} + \frac{u_{y0}}{n^2} \begin{bmatrix} 2c_1 - 2c_2 \\ 4c_3 + 3c_4/2 \\ 2nc_3 \\ -3nc_1 + 4nc_2 \end{bmatrix} \tag{4.81}
$$

where

$$
T_f = n(t_f - t_0), T_i = n(t_i - t_0)
$$
$$
c_1 = 2T_1 - 2T_2 + 2T_3 - T_f, \ c_2 = 2\sin T_1 - 2\sin T_2 + 2\sin T_3 - \sin T_f
$$
$$
c_3 = 2\cos T_1 - 1 - 2\cos T_2 + 2\cos T_3 - \cos T_f, \ c_4 = 2T_1^2 - 2T_2^2 + 2T_3^2 - T_f^2 \tag{4.82}
$$

Solving Eq. (4.81) gives $t_1 \sim t_3$ and t_f, then the optimal control $u_y^*(t)$ and optimal transfer trajectory are obtained.

Optimality is verified as follows. The costate vector $\boldsymbol{\lambda}(t)$ is calculated to derive the curves of switch function $\boldsymbol{\lambda}^T(t)\mathbf{B}$ and the Hamiltonian $H_{\text{time}}(t)$, which further verify whether the minimum conditions and the Hamiltonian are satisfied. The calculation of $\boldsymbol{\lambda}(t)$ is given as follows.

Given $\boldsymbol{\lambda}(t_0) = \boldsymbol{\lambda}_0 = [\lambda_{01} \ \lambda_{02} \ \lambda_{03} \ \lambda_{04}]^T$, Eq. (4.77) derives

$$
\boldsymbol{\lambda}(t) = e^{-\mathbf{A}^T \Delta t}\boldsymbol{\lambda}_0 = \begin{bmatrix} 4 - 3c & 6n\Delta t - 6s & -3ns & 6n(c - 1) \\ 0 & 1 & 0 & 0 \\ -s/n & 2(c - 1)/n & c & 2s \\ 2(1 - c)/n & (3n\Delta t - 4s)/n & -2s & 4c - 3 \end{bmatrix} \begin{bmatrix} \lambda_{01} \\ \lambda_{02} \\ \lambda_{03} \\ \lambda_{04} \end{bmatrix} \tag{4.83}
$$

which can be substituted to Eqs. (4.78) and (4.79) to yield

$$
\begin{bmatrix}
2\cos T_1 - 2 & 4\sin T_1 - 3T_1 & 2n\sin T_1 & n(3 - 4\cos T_1) \\
2\cos T_2 - 2 & 4\sin T_2 - 3T_2 & 2n\sin T_2 & n(3 - 4\cos T_2) \\
2\cos T_3 - 2 & 4\sin T_3 - 3T_3 & 2n\sin T_3 & n(3 - 4\cos T_3) \\
\dot{x}_0 & \dot{y}_0 & 3n^2 x_0 + 2n\dot{y}_0 & -2n\dot{x}_0 + u_{y0}
\end{bmatrix}
\begin{bmatrix}
\lambda_{01} \\ \lambda_{02} \\ \lambda_{03} \\ \lambda_{04}
\end{bmatrix}
=
\begin{bmatrix}
0 \\ 0 \\ 0 \\ -1
\end{bmatrix}
\tag{4.84}
$$

Then λ_0 can be calculated and substituted into Eq. (4.84) to derive $\lambda(t)$.

2. Out-of-plane motion

If $\mathbf{u}(t) = \begin{bmatrix} 0 & 0 & u_z(t) \end{bmatrix}^{\mathrm{T}}$, then

$$
\mathbf{x}_z = \begin{bmatrix} z \\ \dot{z} \end{bmatrix}, \quad
\mathbf{A}_z = \begin{bmatrix} 0 & 1 \\ -n^2 & 0 \end{bmatrix}, \quad
\mathbf{B}_z = \begin{bmatrix} 0 \\ 1 \end{bmatrix}
\tag{4.85}
$$

Bang-Bang control assumes the control curve given as the right one in Fig. 4.9, in which the initial control is $u_{z0} = \pm U_z$. Equation (4.73) can then yield

$$
\begin{cases}
z_f \cos T_{zf} - \dot{z}_f \sin T_{zf}/n = z_0 + [2\cos T_{z1} - (1 + \cos T_{zf})]u_{z0}/n^2 \\
z_f n \sin T_{zf} + \dot{z}_f \cos T_{zf} = \dot{z}_0 + (2\sin T_{z1} - \sin T_{zf})u_{z0}/n
\end{cases}
\tag{4.86}
$$

Further, T_{z1} and T_{zf} can be calculated.

Given $\lambda_z(t_0) = \lambda_{z0} = \begin{bmatrix} \lambda_{z01} & \lambda_{z02} \end{bmatrix}^{\mathrm{T}}$, then

$$
\lambda_z(t) = e^{-\mathbf{A}_z^{\mathrm{T}} \Delta t} \lambda_{z0} =
\begin{bmatrix} c & ns \\ -s/n & c \end{bmatrix}
\begin{bmatrix} \lambda_{z01} \\ \lambda_{z02} \end{bmatrix}
\tag{4.87}
$$

which can be substituted to Eqs. (4.78) and (4.79) to yield

$$
\begin{bmatrix} -\sin T_{z1}/n & \cos T_{z1} \\ -\dot{z}_0 & n^2 z_0 - u_{z0} \end{bmatrix}
\begin{bmatrix} \lambda_{z10} \\ \lambda_{z20} \end{bmatrix}
=
\begin{bmatrix} 0 \\ 1 \end{bmatrix}
\tag{4.88}
$$

Then λ_{z0} can be calculated and substituted into Eq. (4.87) to derive $\lambda_z(t)$. Optimality is verified by generating curves of the switch function and the Hamiltonian.

4.4.3.2 Minimum-Fuel Maneuver

The cost function is

$$
\min J_{\text{fuel}} = \int_{t_0}^{t_f} \|\mathbf{u}(t)\|_1 \, dt
\tag{4.89}
$$

The Hamiltonian is

$$H_{\text{fuel}}(t) = \|\mathbf{u}(t)\|_1 + \boldsymbol{\lambda}^{\mathsf{T}}(t)[\mathbf{A}\mathbf{x}(t) + \mathbf{B}\mathbf{u}(t)] \tag{4.90}$$

The costate equation is

$$\dot{\boldsymbol{\lambda}}(t) = -\mathbf{A}^{\mathsf{T}}\boldsymbol{\lambda}(t) \tag{4.91}$$

The minimum condition is

$$\boldsymbol{\lambda}^{\mathsf{T}}(t_i)\mathbf{B} = -\text{sgn}[\mathbf{u}(t_i)], \ \forall t_i \tag{4.92}$$

where sgn is the sign function. Equation (4.92) implies

$$\mathbf{u}(t) = \begin{cases} |\mathbf{u}| & \boldsymbol{\lambda}^{\mathsf{T}}(t)\mathbf{B} < -1 \\ 0 & -1 \leq \boldsymbol{\lambda}^{\mathsf{T}}(t)\mathbf{B} \leq 1 \\ -|\mathbf{u}| & \boldsymbol{\lambda}^{\mathsf{T}}(t)\mathbf{B} > 1 \end{cases} \tag{4.93}$$

The Hamiltonian satisfies

$$H_{\text{fuel}}(t) = \text{constant}, \ \forall t \tag{4.94}$$

1. In-plane motion
   Equation (4.73) yields

$$\begin{bmatrix} x_f \\ y_f \\ \dot{x}_f \\ \dot{y}_f \end{bmatrix} = \begin{bmatrix} 4 - 3\cos T_f & 0 & \sin T_f/n & 2(1 - \cos T_f)/n \\ 6(\sin T_f - T_f) & 1 & 2(\cos T_f - 1)/n & (4\sin T_f - 3T_f)/n \\ 3n\sin T_f & 0 & \cos T_f & 2\sin T_f \\ 6n(\cos T_f - 1) & 0 & -2\sin T_f & 4\cos T_f - 3 \end{bmatrix}$$

$$\times \left(\begin{bmatrix} x_0 \\ y_0 \\ \dot{x}_0 \\ \dot{y}_0 \end{bmatrix} + \frac{u_{y_0}}{n^2} \begin{bmatrix} 2c_1 - 2c_2 \\ 4c_3 + 3c_4/2 \\ 2nc_3 \\ -3nc_1 + 4nc_2 \end{bmatrix} \right) \tag{4.95}$$

where

$$\begin{aligned}
c_1 &= T_1 + T_2 - T_3 - T_4 + T_5 + T_6 - T_f \\
c_2 &= \sin T_1 + \sin T_2 - \sin T_3 - \sin T_4 + \sin T_5 + \sin T_6 - \sin T_f \\
c_3 &= \cos T_1 + \cos T_2 - \cos T_3 - \cos T_4 + \cos T_5 + \cos T_6 - \cos T_f - 1 \\
c_4 &= T_1^2 + T_2^2 - T_3^2 - T_4^2 + T_5^2 + T_6^2 - T_f^2
\end{aligned}$$

$$\tag{4.96}$$

Combining Eqs. (4.87) and (4.92) yields

$$
\begin{bmatrix}
2\cos T_1 - 2 & 4\sin T_1 - 3T_1 & 2n\sin T_1 & n(3 - 4\cos T_1) \\
2\cos T_2 - 2 & 4\sin T_2 - 3T_2 & 2n\sin T_2 & n(3 - 4\cos T_2) \\
2\cos T_3 - 2 & 4\sin T_3 - 3T_3 & 2n\sin T_3 & n(3 - 4\cos T_3) \\
2\cos T_4 - 2 & 4\sin T_4 - 3T_4 & 2n\sin T_4 & n(3 - 4\cos T_4) \\
2\cos T_5 - 2 & 4\sin T_5 - 3T_5 & 2n\sin T_5 & n(3 - 4\cos T_5) \\
2\cos T_6 - 2 & 4\sin T_6 - 3T_6 & 2n\sin T_6 & n(3 - 4\cos T_6)
\end{bmatrix}
\begin{bmatrix}
\lambda_{01} \\ \lambda_{02} \\ \lambda_{03} \\ \lambda_{04}
\end{bmatrix}
= n\mathrm{sgn}(u_0)
\begin{bmatrix}
1 \\ -1 \\ -1 \\ 1 \\ 1 \\ -1
\end{bmatrix}
$$

(4.97)

In Eqs. (4.95) and (4.97), there are altogether ten equations and ten unknowns, which can be used to solve T_i and λ_0. The solution procedures are presented as follows. First, four equations in Eq. (4.97) are selected to express λ_0 as a function of T_i. Second, substitute T_i into the remaining two equations in Eq. (4.97) to form six equations with Eq. (4.95). Third, solve these equations to yield T_i, which further derives λ_0. Finally, λ_0 is substituted into Eq. (4.87) to get $\lambda(t)$. Optimality is verified by generating curves of the switch function and the Hamiltonian.

2. Out-of-plane motion

Equation (4.73) yields

$$
\begin{cases}
z_f \cos T_{zf} - \dot{z}_f \sin T_{zf}/n = z_0 + (\cos T_{z2} + \cos T_{z3} - \cos T_{z1} - \cos T_{z4})u_{z0}/n^2 \\
z_f n \sin T_{zf} + \dot{z}_f \cos T_{zf} = \dot{z}_0 + (\sin T_{z1} + \sin T_{z2} - \sin T_{z1} - \sin T_{z4})u_{z0}/n
\end{cases}
$$

(4.98)

Substituting Eq. (4.87) into Eq. (4.92) yields

$$
\begin{bmatrix}
-\sin T_{z1}/n & \cos T_{z1} \\
-\sin T_{z1}/n & \cos T_{z1} \\
-\sin T_{z1}/n & \cos T_{z1} \\
-\sin T_{z1}/n & \cos T_{z1}
\end{bmatrix}
\begin{bmatrix}
\lambda_{z01} \\ \lambda_{z02}
\end{bmatrix}
= -\mathrm{sgn}(u_{z0})
\begin{bmatrix}
1 \\ 1 \\ -1 \\ -1
\end{bmatrix}
$$

(4.99)

Equations (4.98) and (4.99) have altogether six equations and six unknowns, which can be calculated to derive $t_{z1} \sim t_{z4}$ and λ_{z0}. Then λ_{z0} is substituted into Eq. (4.87) to derive $\lambda_z(t)$. Optimality is verified by generating curves of the switch function and the Hamiltonian.

4.4.4 Numerical Simulation

Assume the customer runs in a circular orbit of 1,000 km.

$$
t_0 = 0, \mathbf{r}_0 = \begin{bmatrix} 180 & -350 & 100 \end{bmatrix}^\mathrm{T} \mathrm{m}, \mathbf{v}_0 = \begin{bmatrix} -0.15 & -0.40 & -0.06 \end{bmatrix}^\mathrm{T} \mathrm{m/s};
$$

$$
\mathbf{r}_f = \begin{bmatrix} 0 & 200 & 0 \end{bmatrix}^\mathrm{T} \mathrm{m}, \mathbf{v}_f = \begin{bmatrix} 0 & 0 & 0 \end{bmatrix}^\mathrm{T} \mathrm{m/s};
$$

$$
U_y = U_z = 6.0 \times 10^{-5} \, \mathrm{m/s}.
$$

Fig. 4.10 In-plane minimum-time maneuver trajectory

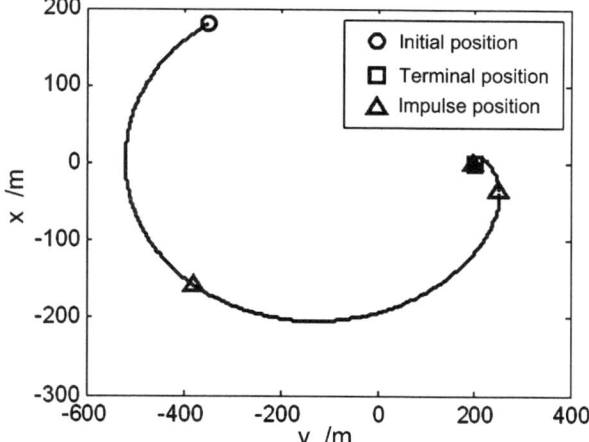

Fig. 4.11 Switch function of in-plane minimum-time maneuver

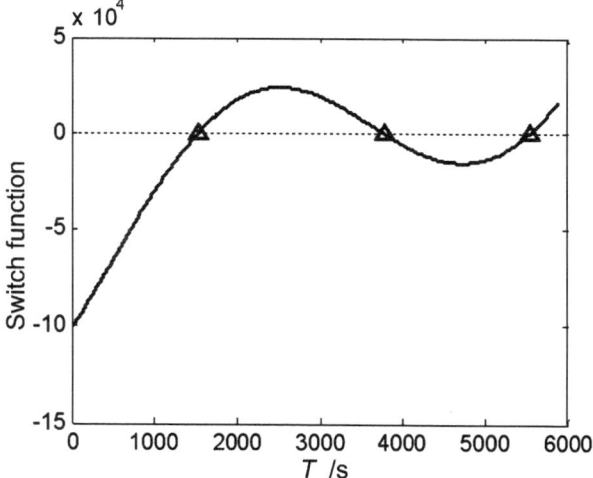

The in-plane minimum-time transfer trajectory is shown in Fig. 4.10, $t_f = 5,894$ s, $u_{y0} = U_y$, the switch times $t_1 \sim t_3$ are 1,530 s, 3,777 s, and 5,538 s respectively, $\Delta v_y = 0.3536$ m/s.

The switch function curve is given as in Fig. 4.11, in which "$\triangle$" denotes switch points. The Hamiltonian curve is given as in Fig. 4.12, verifying the optimality.

Simulation results of the out-of-plane minimum-time maneuver are $t_{zf} = 3,009$ s, $u_{z0} = U_z$, $t_{z1} = 2,545$ s, and $\Delta v_z = 0.1805$ m/s.

It is obvious that if the minimum-time maneuver takes $\Delta t_{min} = 5,894$ s, the minimum-fuel maneuver takes $t_f > 5,894$ s. Given $t_f = 7,200$ s, the minimum-fuel transfer trajectory is shown as in Fig. 4.13. And $u_{y0} = U_y$, meanwhile the corresponding switch times $t_1 \sim t_6$ are 1,022 s, 2,036 s, 3,241 s, 4,679 s, 5,866 s, and 6,884 s respectively, $\Delta v_y = 0.2238$ m/s. The switch function curve is given as in

Fig. 4.12 The Hamiltonian
of in-plane minimum-time
maneuver

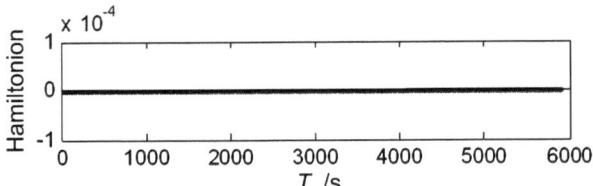

Fig. 4.13 In-plane
minimum-fuel maneuver
trajectory

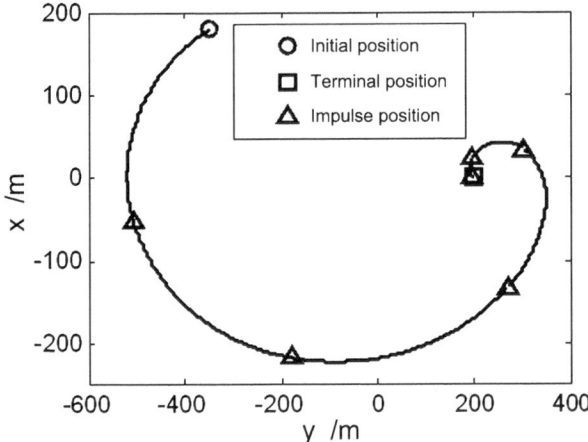

Fig. 4.14 Switch function
of in-plane minimum-fuel
maneuver

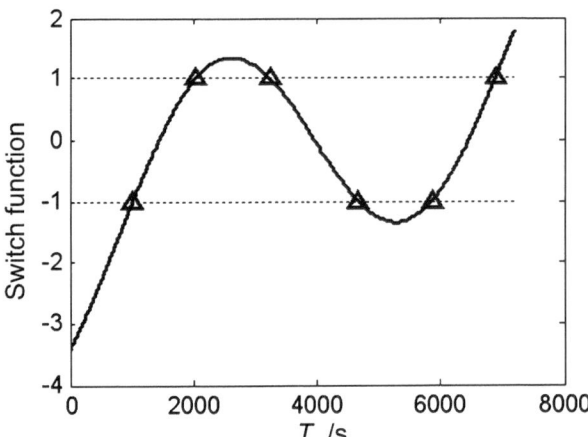

Fig. 4.14, and the Hamiltonian curve is given as in Fig. 4.15, both verifying the optimality.

The switch times $t_{z1} \sim t_{z4}$ of out-of-plane minimum-fuel maneuver are 527 s, 1,538 s, 3,681 s, and 4,692 s respectively, $u_{z0} = U_z$, and $\Delta v_z = 0.1214$ m/s.

Finally, the minimum-fuel maneuver trajectory of a fixed time mission ($t_f = 7,200$ s) is depicted as in Fig. 4.16. $\Delta V = 0.3458$ m/s.

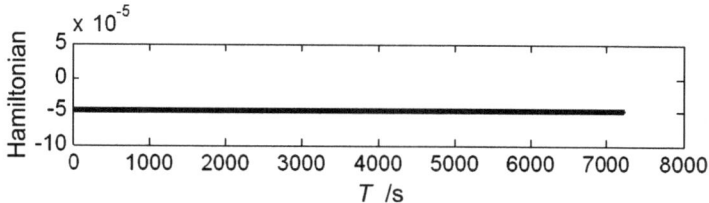

Fig. 4.15 The Hamiltonian of in-plane minimum-fuel maneuver

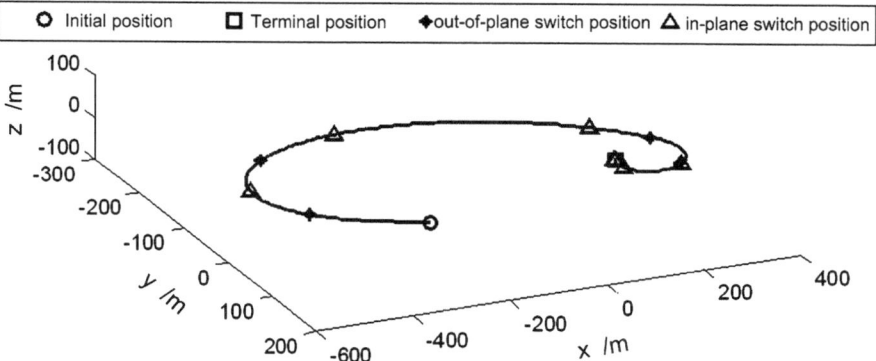

Fig. 4.16 Minimum-fuel maneuver trajectory of a fixed time mission

4.5 *hp*-APM for Local Inspection Trajectory Planning

Spacecraft proximity inspection, which is categorized as global and local inspection, is a crucial phase in the on-orbit operations mission. Global inspection usually assumes a natural or constrained periodic trajectory, while local inspection can only be accomplished by a constrained trajectory, which is more complex. Therefore, the local inspection trajectory planning is the focus of the following sections.

4.5.1 Mission Formulation

The local inspection mission scenario is presented as in Fig. 4.17. The servicer whose trajectory and attitude control are enabled by thrusters must ensure its sensor can always capture the customer while avoid getting into the customer FOV. Meanwhile, safety constraint and imaging quality requirement must be satisfied [10, 11]. In summary, the mission requirements could be formulated by the following constraints.

Fig. 4.17 Concealment
and safety constraint

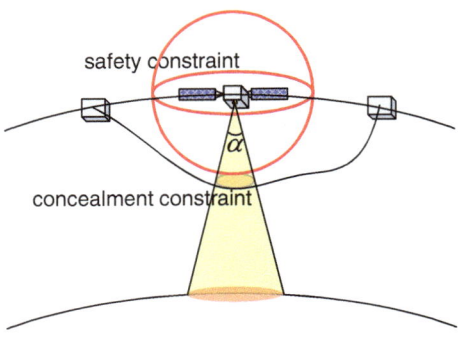

Fig. 4.18 Attitude
constraint

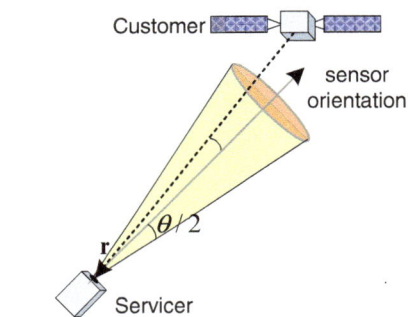

1. *Concealment constraint*

   As shown in Fig. 4.17, the concealment constraint must satisfy

   $$\cos^{-1}\left[\mathbf{r} \cdot \mathbf{R}_c^o \mathbf{N}_c^{\mathrm{view}}/|\mathbf{r}|\right] > \alpha/2 \tag{4.100}$$

   where α is customer FOV, $\mathbf{N}_c^{\mathrm{view}}$ customer sensor orientation in the customer body frame, and $\mathbf{R}_c^o$ rotation matrix from the customer body frame to Hill frame.

2. *Attitude constraint*

   As shown in Fig. 4.18, the attitude constraint must satisfy

   $$\cos^{-1}\left[(-\mathbf{r}) \cdot \mathbf{R}_s^o \mathbf{N}_s^{\mathrm{view}}/|\mathbf{r}|\right] \le \theta/2 \tag{4.101}$$

   where θ is servicer FOV, $\mathbf{N}_s^{\mathrm{view}}$ servicer sensor orientation in the servicer body frame, and $\mathbf{R}_s^o$ rotation matrix from the servicer body frame to Hill frame.

3. *Inspection angle constraint*

   As shown in Fig. 4.19, for better imaging quality, the inspection angle σ must satisfy

Fig. 4.19 Inspection angle
constraint

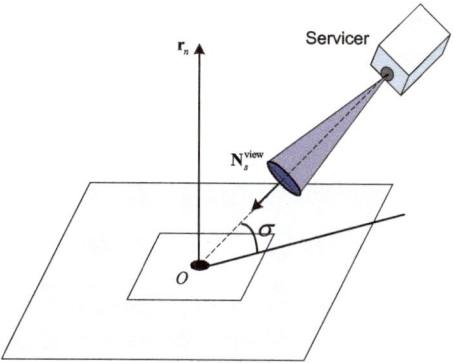

$$\sigma = \frac{\pi}{2} - \cos^{-1}\left(\mathbf{R}_s^o \frac{-\mathbf{N}_s^{\text{view}}}{\left|\mathbf{N}_s^{\text{view}}\right|} \cdot \mathbf{R}_c^o \frac{\mathbf{r}_n}{\left|\mathbf{r}_n\right|} \right) \geq \sigma_{\min} \qquad (4.102)$$

where $\mathbf{r}_n$ is normal vector to the customer area and $\sigma_{\min}$ minimum inspection angle.

4. *Safety constraint*

The servicer should always keep a safe distance r_{safe} from the customer, written as

$$|\mathbf{r}| \geq r_{\text{safe}} \qquad (4.103)$$

5. *Control constraint*

Assume continuous low thrusters are applied, then

$$|\mathbf{f}_s(t)| = \begin{bmatrix} |f_{sx}(t)| & |f_{sy}(t)| & |f_{sz}(t)| \end{bmatrix}^{\text{T}} \leq \mathbf{f}_{\max} \qquad (4.104)$$

where $\mathbf{f}_{\max}$ is the upper boundary of control acceleration.

6. *Resolution constraint*

The relationship between inspection distance and image resolution is shown as

$$\text{Res} = r \tan\left(\theta/P_{\text{num}}\right) \approx r\theta/P_{\text{num}} \qquad (4.105)$$

where r is inspection distance and P_{num} pixels.

To ensure image resolution, the inspection distance must satisfy

$$r \leq \text{Res} \cdot P_{\text{num}}/\theta \qquad (4.106)$$

7. *Pixel smear constraint*

Pixel smear is defined by

$$\text{Ps} = P_{\text{num}} \operatorname{atan}\left(vt_{\text{exp}}/r\right)/\theta \approx vt_{\text{exp}}P_{\text{num}}/(r\theta) \qquad (4.107)$$

where v is relative velocity perpendicular to $\mathbf{r}$ and t_{exp} exposure time.

Then, to reduce pixel smear, v must satisfy

$$v \leq \text{Ps} \cdot r\theta/\left(t_{\text{exp}}P_{\text{num}}\right) \qquad (4.108)$$

4.5.2   6-DOF Coupled Dynamic Model

If the customer runs in a near-circular orbit, the relative trajectory dynamic model is formulated as Eq. (4.3), (see Sect. 4.2.2.1).

Neglecting disturbance torques, the attitude dynamics of the servicer in its body frame is given as

$$\mathbf{J}\dot{\boldsymbol{\omega}} + \boldsymbol{\omega} \times \mathbf{J}\boldsymbol{\omega} = \mathbf{T} \tag{4.109}$$

where $\mathbf{J}$ is inertial matrix, $\boldsymbol{\omega}$ absolute angular velocity, and $\mathbf{T}$ control torque.

The attitude kinematics model with quaternions is presented as

$$\dot{\mathbf{q}} = \frac{1}{2}\boldsymbol{\Omega}(\boldsymbol{\omega}^o)\mathbf{q} \tag{4.110}$$

where $\mathbf{q} = [q_1, q_2, q_3, q_4]^T$, and $\boldsymbol{\Omega}(\boldsymbol{\omega}^o)$ is defined as

$$\boldsymbol{\Omega}(\omega^o) = \begin{bmatrix} 0 & \omega_z^o & -\omega_y^o & \omega_x^o \\ -\omega_z^o & 0 & \omega_x^o & \omega_y^o \\ \omega_y^o & -\omega_x^o & 0 & \omega_z^o \\ -\omega_x^o & -\omega_y^o & -\omega_z^o & 0 \end{bmatrix} \tag{4.111}$$

where $\boldsymbol{\omega}^o$ is angular velocity with respect to Hill frame.

$$\left[\omega_x^o, \omega_y^o, \omega_z^o\right]^T = \mathbf{R}_s^o\left[\omega_x, \omega_y, \omega_z\right]^T - [0, 0, n]^T \tag{4.112}$$

where $\mathbf{R}_s^o$ is written with quaternions as

$$\mathbf{R}_s^o = \begin{bmatrix} q_4^2 + q_1^2 - q_2^2 - q_3^2 & 2(q_1q_2 - q_3q_4) & 2(q_1q_3 + q_2q_4) \\ 2(q_1q_2 + q_3q_4) & q_4^2 - q_1^2 + q_2^2 - q_3^2 & 2(q_2q_3 + q_1q_4) \\ 2(q_1q_3 - q_2q_4) & 2(q_2q_3 + q_1q_4) & q_4^2 - q_1^2 - q_2^2 + q_3^2 \end{bmatrix} \tag{4.113}$$

Assume the servicer can be approximated as a cuboid $L_1 \times L_2 \times L_3$, and the six continuous low thrusters are configured as Fig. 4.20 [12], then the input matrix of thrusters in the servicer body frame can be presented as

$$\mathbf{B}_{in} = \begin{bmatrix} 1 & -1 & 0 & 0 & 0 & 0 \\ 0 & 0 & 1 & -1 & 0 & 0 \\ 0 & 0 & 0 & 0 & 1 & -1 \\ 0 & 0 & L_3/2 & L_3/2 & L_2/2 & L_2/2 \\ -L_3/2 & -L_3/2 & 0 & 0 & -L_1/2 & -L_1/2 \\ L_2/2 & L_2/2 & L_1/2 & L_1/2 & 0 & 0 \end{bmatrix} \tag{4.114}$$

Fig. 4.20 Configuration
of thrusters

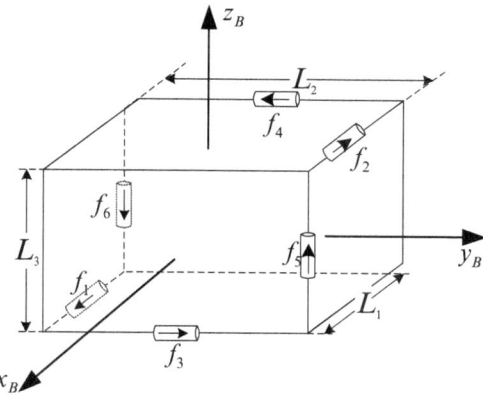

Define $\mathbf{f} = [f_1 \quad f_2 \quad f_3 \quad f_4 \quad f_5 \quad f_6]^{\mathrm{T}}$, then

$$
\begin{bmatrix} \mathbf{u} \\ \mathbf{T} \end{bmatrix} = \begin{bmatrix} \mathbf{R}_s^o & \mathbf{0}_{3\times3} \\ \mathbf{0}_{3\times3} & \mathbf{I}_{3\times3} \end{bmatrix} \mathbf{B}_{\mathrm{in}} \mathbf{f} \tag{4.115}
$$

Combining Eqs. (4.3), (4.109), and (4.115), the coupled dynamics is given as

$$
\begin{bmatrix} \dot{\mathbf{x}} \\ \dot{\mathbf{q}} \\ \dot{\boldsymbol{\omega}} \end{bmatrix} = \begin{bmatrix} \mathbf{Ax} \\ -\frac{1}{2}\boldsymbol{\Omega}(\omega^o)\mathbf{q} \\ -\mathbf{J}^{-1}(\boldsymbol{\omega}\times\mathbf{J}\boldsymbol{\omega}) \end{bmatrix}_{13\times1} + \begin{bmatrix} \mathbf{B}_{6\times3}\mathbf{R}_s^o & \mathbf{0}_{6\times3} \\ \mathbf{0}_{4\times3} & \mathbf{0}_{4\times3} \\ \mathbf{0}_{3\times3} & \mathbf{J}_{3\times3}^{-1} \end{bmatrix}_{13\times6} (\mathbf{B}_{\mathrm{in}}\mathbf{f})_{6\times1} \tag{4.116}
$$

4.5.3 *Planning Model*

The fuel-optimal cost function is defined as

$$
J = \int_{t_0}^{t_f} \left(f_1^2 + f_2^2 + f_3^2 + f_4^2 + f_5^2 + f_6^2\right) \mathrm{d}t \tag{4.117}
$$

The decision variable is $\mathbf{f}$, the constraints are listed as Eqs. (4.100–4.108), and the dynamic model is shown as Eq. (4.116). Therefore, the planning model is nonconvex and nonlinear, which is difficult to solve. In the following sections, we attempt to adopt *hp*-APM.

4.5.4 *hp-APM*

The *hp*-APM [13, 14] was put forward by Darby et al. to offset the limitations of conventional pseudospectral methods, such as slow convergence. This method utilizes multi-resolution technology to divide the trajectory into segments and allow the number of segments, segment widths, and polynomial degrees to vary throughout the time interval of interest. It is viable for efficiently and accurately solving complex optimization control problems.

First, *hp*-APM selects collocation points randomly and obtains the error vectors of the midpoint of two adjacent collocation points by polynomial interpolation. Second, the error vectors are used to determine which segments need to increase collocation points and which segments need to be further divided until precision requirements are satisfied.

Divide the trajectory into S segments. For each segment s ($s = 1, \ldots, S$), denote N_{cp} as the number of collocation points, and $[t_{s-1}, t_s]$ as the time span.

Define $\mathbf{X}(\bar{t}_i)$ and $\mathbf{u}(\bar{t}_i)$ as the state and control vector of the midpoint of two adjacent collocation points by polynomial interpolation, then

$$
\begin{aligned}
\overline{\mathbf{X}} &= \begin{bmatrix} \mathbf{X}(\bar{t}_1) & \cdots & \mathbf{X}(\bar{t}_{N_{\text{cp}}-1}) \end{bmatrix}^{\mathrm{T}}_{(N_{\text{cp}}-1)\times l} \\
\overline{\mathbf{U}} &= \begin{bmatrix} \mathbf{u}(\bar{t}_1) & \cdots & \mathbf{u}(\bar{t}_{N_{\text{cp}}-1}) \end{bmatrix}^{\mathrm{T}}_{(N_{\text{cp}}-1)\times m}
\end{aligned}
\tag{4.118}
$$

where l refers to state dimension, m control dimension, and

$$
\bar{t}_i = \frac{t_i + t_{i+1}}{2} \quad i = 1, \ldots, N_{\text{cp}} - 1
\tag{4.119}
$$

1. Principle for collocation points increase or segment division based on dynamics constraint error

   According to *hp*-APM, $[t_{s-1}, t_s]$ is mapped to $[-1, 1]$ and the corresponding $\bar{t}_i$ is denoted as $\bar{\tau}_i$. The dynamics constraint at the midpoints could be written as [13]

$$
\sum_{i=1}^{N_{\text{cp}}-1} \overline{\mathbf{D}}_{ki}\overline{\mathbf{X}}_i - \frac{t_s - t_{s-1}}{2}\mathbf{F}\left(\overline{\mathbf{X}}_k, \overline{\mathbf{U}}_k, \bar{\boldsymbol{\tau}}_k, t_{s-1}, t_s\right) = 0 \quad k = 1, \ldots, N_{\text{cp}} - 1
\tag{4.120}
$$

where $\overline{\mathbf{D}}$ is differentiation matrix, and $\bar{\boldsymbol{\tau}}_k = \begin{bmatrix} \bar{\tau}_1, \ldots, \bar{\tau}_{N_{\text{cp}}-1} \end{bmatrix}^{\mathrm{T}}$.

The error matrix of dynamics constraint is

$$
\mathbf{E} = \left| \overline{\mathbf{D}}\overline{\mathbf{X}} - \frac{t_s - t_{s-1}}{2}\mathbf{F}\left(, \overline{\mathbf{X}}, \overline{\mathbf{U}}, \bar{\boldsymbol{\tau}}, t_{s-1}, t_s\right) \right| \in \mathbb{R}^{(N_{\text{cp}}-1)\times l}
\tag{4.121}
$$

Defining $\mathbf{e} = \begin{bmatrix} e(\bar{\tau}_1), \ldots, e(\bar{\tau}_{N_{\text{cp}}-1}) \end{bmatrix}^{\mathrm{T}}$ as the column vector that contains the maximum element of $\mathbf{E}$, and $\bar{e}$ the arithmetic mean of the components of $\mathbf{e}$, the scaled error vector $\overline{\mathbf{e}}$ is formulated as

$$
\overline{\mathbf{e}} = \mathbf{e}/\bar{e}
\tag{4.122}
$$

Introduce $\varepsilon > 0$ and $\delta > 1$ as the user-defined tolerance, then if max $\mathbf{E} > \varepsilon$ and $\forall \bar{e}(\bar{\tau}_i) \leq \delta$, collocation points should be increased as

$$N_{\mathrm{cp}}^{(j+1)} = N_{\mathrm{cp}}^{(j)} + \frac{N_{\mathrm{cp}}^{(j)} - N_{\mathrm{cp}}^{(j-1)}}{m^{(j)} - m^{(j-1)}} \left(|\lg(\varepsilon)| - m^{(j)} \right) \qquad (4.123)$$

where $N_{\mathrm{cp}}^{(j)}$ is the number of collocation points on grid iteration j, $m^{(j)}$ the error order by grid iteration j, i.e., $O\left(10^{-m^{(j)}}\right)$.

If max $\mathbf{E} > \varepsilon$ and $\exists \bar{e}(\bar{\tau}_i) > \delta$, the segment needs to be further divided at $\bar{e}(\bar{\tau}_{\max}) = \max \bar{e}$, where $\bar{\tau}_{\max}$ is the place to put the break.
2. Principle for segment division based on trajectory constraint error
   The trajectory constraint is expressed as

$$\mathbf{C}\left(\overline{\mathbf{X}}(\bar{\tau}_i), \overline{\mathbf{u}}(\bar{\tau}_i); t_{s-1}, t_s\right) \leq 0 \quad k = 1, 2, \ldots, N_{\mathrm{cp}-1} \qquad (4.124)$$

then the error vector of trajectory constraint of the midpoint of two adjacent collocation points is written as

$$\mathbf{e}\left(\bar{\tau}_{\max}'\right) = \begin{cases} 0 & \text{if } \mathbf{C}\left(\overline{\mathbf{X}}(\bar{\tau}_i), \overline{\mathbf{u}}(\bar{\tau}_i); t_{s-1}, t_s\right) \leq 0 \\ \ell \max \mathbf{C}\left(\overline{\mathbf{X}}(\bar{\tau}_i), \overline{\mathbf{u}}(\bar{\tau}_i); t_{s-1}, t_s\right) & \text{otherwise} \end{cases}$$
$$(4.125)$$

where ℓ is the normalized coefficient, which differs with trajectories.

If $\mathbf{e}\left(\bar{\tau}_{\max}'\right) > \varepsilon$, segment s is divided at $\bar{\tau}_{\max}'$.

The iterative procedures are summarized as in Fig. 4.21

4.5.5 Numerical Simulation

Assume the customer is a GEO satellite; the angle of the FOV is $30°$ and the safety zone is a customer-centered sphere with a radius of 200 m. The mission duration is 5,000 s. For the servicer, suppose the initial and terminal value in Eq. (4.116) as

$$[0, -500 \text{ m}, 0, 0, 0, 0, 0, 0, 0, 1]^{\mathrm{T}} \text{ and}$$
$$[0, 500 \text{ m}, 0, 0, 0, 0, 0, 0, 0, -1, 0, 0, 0]^{\mathrm{T}}.$$

Simulation parameters are set as follows.

$$f_{\max} = 0.002 \text{ m/s}^2; \, l_1 \times l_2 \times l_3 = 0.2 \text{ m} \times 0.2 \text{ m} \times 0.2 \text{ m}; \, \sigma_{\min} = 5°,$$
$$t_{\exp} = 30 \text{ ms and } P_{\mathrm{num}} = 1024, \text{ Ps} < 1, \text{ Res} \geq 5 \text{ cm}, S_0 = 1, N_{\mathrm{cp}} = 15.$$

The planning model solved by *hp*-APM could be addressed by the Open-source pseudospectral optimization control software GPOPS using the NLP solver SNOPT [14]. The simulation is carried out by a computer with 2.8GHZ CPU and 2.0G

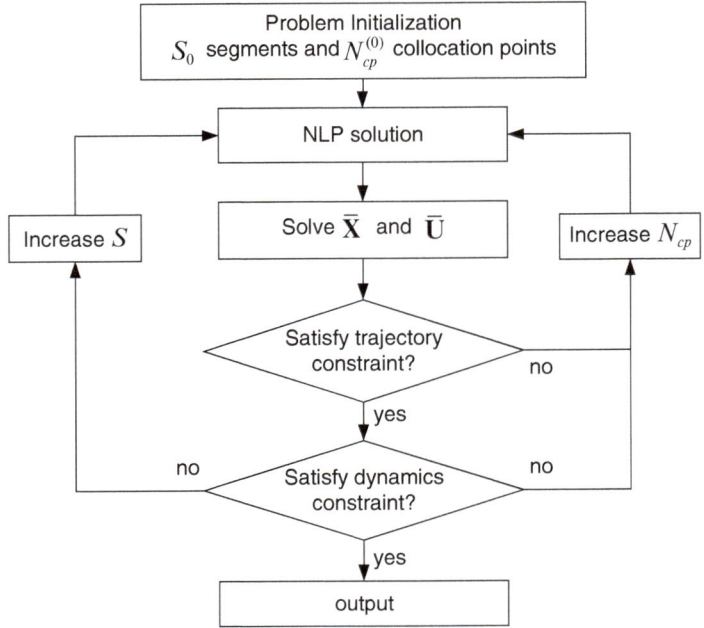

Fig. 4.21 *hp*-APM procedure

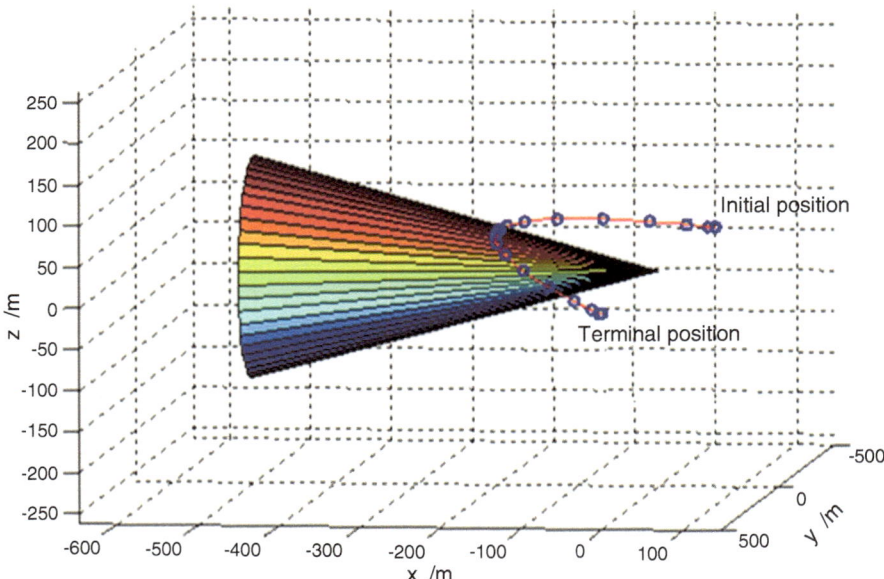

Fig. 4.22 The inspection trajectory of *hp*-APM

RAM. Suppose the number of collocation points in the segment that needs to be further divided is 10. If the computation precision is 10^{-1}, the inspection trajectory can be generated within 20 s, while if the precision is 10^{-2}, the generation time would be approximately 30 s, verifying the high efficiency of *hp*-APM.

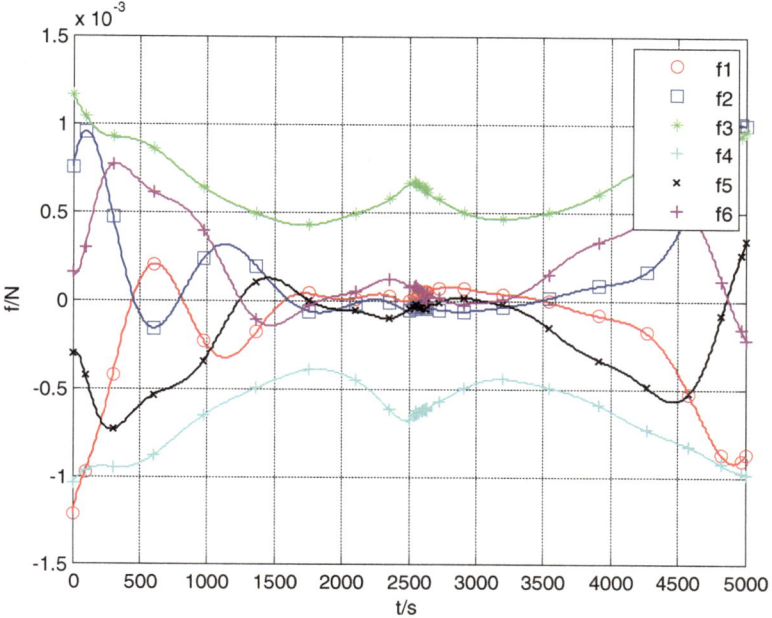

Fig. 4.23 The thruster force curve of *hp*-APM

The inspection trajectory and corresponding control curve are depicted in Figs. 4.22 and 4.23.

In Fig. 4.22, the blue circles denote the collocation points, whose distribution could be optimized to achieve hp-APM efficiency. In Fig. 4.23, the control curves are smooth and the magnitude is no more than 1.5×10^{-3} m/s^2, indicating engineering possibility.

4.6   IAPF for Close Proximity Inspection

When the proximity inspection mission is performed on a complicatedly structured spacecraft (e.g., International Space Station), the mutual shading effects would make blind zones a problem. Therefore, the inspection trajectory is required to be as close as possible to the customer for scanning its surface [15, 16], which is named close proximity inspection.

4.6.1   Mission Formulation

In close proximity operations, the key is to generate a trajectory which can follow the customer surface well.

Fig. 4.24 Inspection space

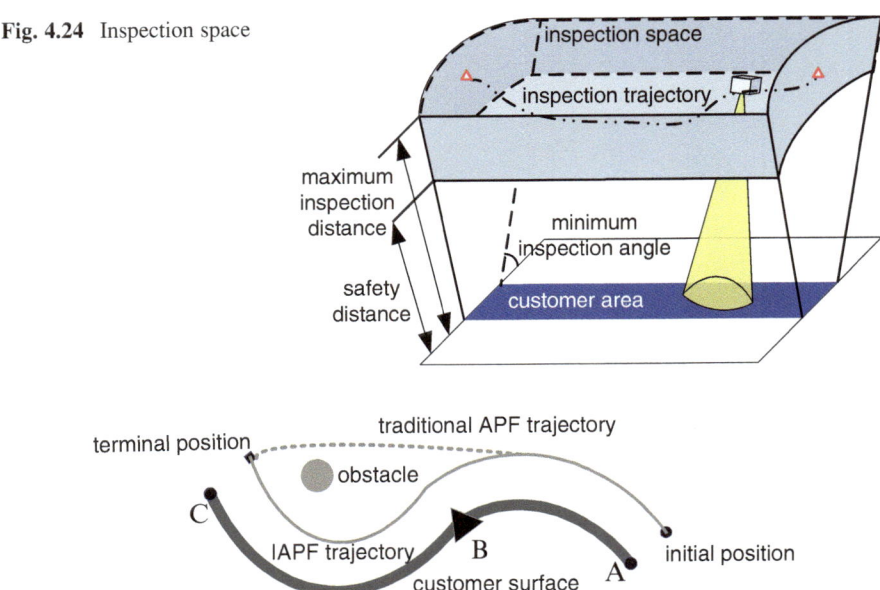

Fig. 4.25 A comparison between traditional APF and IAPF

As shown in Fig. 4.24, the inspection trajectory must be within a fan-shaped inspection space, whose outer edge is defined by the maximum inspection distance, inner edge by the collision avoidance constraint, and left and right edges by the minimum angle of view constraint. So the inspection trajectory must be generated online, which requires an efficient planning algorithm such as IAPF.

4.6.2 IAPF Algorithm

As shown in Fig. 4.25, the servicer fails to follow the customer surface by traditional APF, which highlights the need to design a new trajectory planning algorithm [17–19].

Making use of the follow-wall behavior of the terrestrial robot, the acceleration which is tangent to the customer surface and points to the terminal position by the attractive potential enables the servicer to follow the customer surface well. The relative velocity is determined by the size of the potential field to satisfy the pixel smear constraint. The acceleration offered by the repulsive potential field keeps the servicer in the inspection area to satisfy the maximum inspection distance constraint, minimum angle of view constraint, and obstacle avoidance constraint. The total acceleration is the sum of attractive acceleration and repulsive acceleration, written as

$$\mathbf{a} = \mathbf{a}_g + \mathbf{a}_o + \mathbf{a}_d + \mathbf{a}_v \qquad (4.126)$$

where $\mathbf{a}_g$ is attractive acceleration, $\mathbf{a}_o$, $\mathbf{a}_d$, $\mathbf{a}_v$ repulsive accelerations for obstacle avoidance, maximum inspection distance, and minimum angle of view respectively.

Define $\mathbf{v}_d$ as the desired relative velocity, and $\mathbf{v}_d = \mathbf{v}_{gd} + \mathbf{v}_{od} + \mathbf{v}_{dd} + \mathbf{v}_{vd}$.

1. Attractive acceleration

The attractive potential field function is designed as

$$V_g = \frac{\lambda_g}{2}(\mathbf{C}(\phi)\mathbf{r}_{ct})^{\mathrm{T}}(\mathbf{C}(\phi)\mathbf{r}_{ct}) \tag{4.127}$$

where $\mathbf{r}_{ct}$ is the vector from the current to terminal position, $\lambda_g \geq 0$ proportional coefficient (set $\lambda_g = 1/r_{ct}$), ϕ the angle formed by $\mathbf{r}_{ct}$ and the plane tangent to the inspection point, and $\mathbf{C}(\phi)$ the orientation transformation matrix from $\mathbf{r}_{ct}$ to the tangent plane.

Consequently, $\mathbf{a}_g$ can be expressed as

$$\mathbf{a}_g = (\mathbf{v}_{gd} - \mathbf{v})/\Delta t \tag{4.128}$$

where $\mathbf{v}$ is the current relative velocity, and $\mathbf{v}_{gd}$ can be expressed as

$$\mathbf{v}_{gd} = -k_g v_{max}(1 - e^{-t})\left(1 - e^{-(b_g V_g)}\right)\frac{\nabla V_g}{|\nabla V_g|} \tag{4.129}$$

where ∇V_g is the gradient of the attractive potential field, v_{max} the maximum relative velocity, $k_g \in [0, 1]$ and b_g proportional coefficients.

2. Repulsive acceleration

Define $\mathbf{r}_{co}$ as the vector from the current to the obstacle position, and

$$D_o = d_o\left(r_{safe} + v^2/(2a_{max})\right) \tag{4.130}$$

where $d_o \geq 1$ is proportional coefficient. If $r_{co} \leq D_o$, $\mathbf{a}_o$ can be written as

$$\mathbf{a}_o = (\mathbf{v}_{od} - \mathbf{v}_{co})/\Delta t \tag{4.131}$$

where $\mathbf{v}_{co}$ is the projection of $\mathbf{v}$ in the direction of $\mathbf{r}_{co}$, and $\mathbf{v}_{od}$ can be expressed as

$$\mathbf{v}_{od} = k_o v_{max}\left(1 - e^{-b_o(D_o - r_{co})}\right)\frac{\mathbf{r}_{co}}{r_{co}} \tag{4.132}$$

where $k_o \in [0, 1]$ and b_o are proportional coefficients.

Define

$$D_d = d_d\left(D_{max} - v^2/(2a_{max})\right) \tag{4.133}$$

where $d_d \leq 1$ is proportional coefficient.

If $r_{co} \geq D_d$, $\mathbf{a}_d$ can be written as

$$\mathbf{a}_d = (\mathbf{v}_{dd} - \mathbf{v}_{co})/\Delta t \tag{4.134}$$

where

$$\mathbf{v}_{dd} = -k_d v_{max}\left(1 - e^{-b_d(r_{co}-D_d)}\right)\frac{\mathbf{r}_{co}}{r_{co}} \tag{4.135}$$

where $k_d \in [0, 1]$ and b_d are proportional coefficients.

Define

$$\Delta_v = d_v\left(\sigma_{min} + v^2/(2r_{co}a_{max})\right) \tag{4.136}$$

where $d_v \geq 1$ is proportional coefficient.

If $\sigma \leq \Delta_v$, $\mathbf{a}_v$ can be written as

$$\mathbf{a}_v = \frac{\mathbf{v}_{dd} - \mathbf{v}_v}{\Delta t}\frac{\mathbf{U}_v}{U_v} \tag{4.137}$$

where $\mathbf{v}_v$ is the projection of $\mathbf{v}$ in the direction of control acceleration $\mathbf{U}_v$, and

$$\mathbf{v}_v = -k_v v_{max}\left(1 - e^{-b_v(\sigma-\Delta_v)}\right)\frac{\mathbf{U}_v}{U_v} \tag{4.138}$$

$$\mathbf{U}_v = \mathbf{r}_n - \mathbf{r}_n\frac{\mathbf{r}_{co}}{r_{co}}\cos\sigma \tag{4.139}$$

where $k_v \in [0, 1]$ and b_v are proportional coefficients.

4.6.3 Control Parameter Optimization

The IAPF-based online trajectory planning involves several user-defined control parameters, such as D_{max} and v_{max}, which determine the optimized trajectory. Given the customer structure, the control parameters can be optimized by the following multi-objective planning model.

The optimization criteria are selected as trajectory following capacity and control acceleration smooth property, written as

$$\begin{aligned}
\min f_1 &= \sqrt{(r_{co} - \bar{r}_{co})^2} \\
\min f_2 &= \sum (\|a(t_{i+1}) - a(t_i)\|)
\end{aligned} \tag{4.140}$$

where $\bar{r}_{co}$ is the average distance between inspection trajectory and customer surface and t_i the sampling moment of control acceleration.

Fig. 4.26 Profile
of inspection space

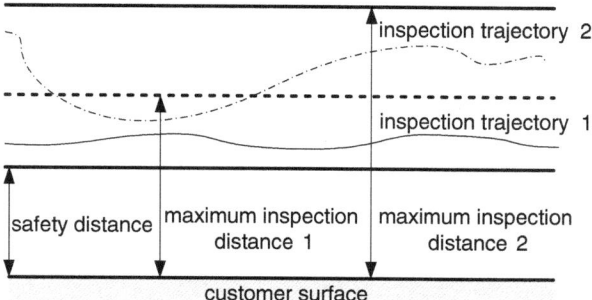

As shown in Fig. 4.26, the smaller the D_{max} is, the better the trajectory following capacity. So f_1 is mainly determined by D_{max}. The main influencing factors of f_2 include v_{max}, k_o, k_{cm}, b_o, b_{cm}, and b_{ca}. To improve solution efficiency, optimization is only performed on critical ones.

Then, the decision variable is

$$\begin{bmatrix} D_{max} & v_{max} & k_o & k_{cm} \end{bmatrix}^T \tag{4.141}$$

which can be solved by the Multi-Objective Evolutionary Algorithm based on Decomposition (MOEA/D), which was first put forward by Zhang [20]. MOEA/D, essentially combining traditional mathematical planning approaches with the evolutionary algorithm, decomposes an MOOP to several subproblems of scalar optimization and further solves these single-objective problems by the evolutionary algorithm.

4.6.4 Numerical Simulation

Assume the customer is in a circular orbit of 450 km. The servicer FOV is $30° \times 30°$ and the initial relative velocity is 0. Other simulation parameters are

$$f_{max} = 0.001 \text{ m/s}^2, t_{exp} = 40 \text{ ms}, P_{num} = 1024, \text{Ps} < 1, r_{safe} = 3.5 \text{ m}.$$

In this simulation, the trajectories of scanning inspection along two different customer surfaces are obtained by the IAPF algorithm, as shown in Figs. 4.27 and 4.28, where the initial and terminal positions are represented by "circle" and "square" respectively.

In Fig. 4.27, the inspection mission takes 3,975.3 s, and Δv is 0.3102 m/s. In Fig. 4.28, it takes 6,920.0 s, and Δv is 1.0981 m/s. Simulation results show the servicer could follow customers with different geometric features.

Suppose the customer surface is shown as Fig. 4.29 and the Pareto optimal solution set of Eq. (4.141) is shown as Fig. 4.30.

Select $\begin{bmatrix} D_{max} & v_{max} & k_o & k_{cm} \end{bmatrix}^T = [5.846, 0.02, 1.059, 1.0718]^T$, then $f_1 = 0.9637$ m, $f_2 = 2.5443$ m/s^2. The inspection trajectory is shown as in Fig. 4.29. The control acceleration curves before and after optimization are given as in Fig. 4.31.

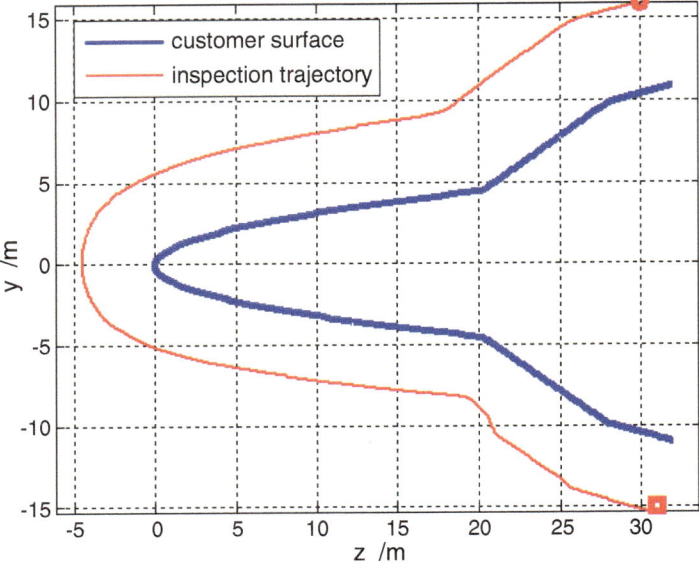

Fig. 4.27 The inspection trajectory of case-1

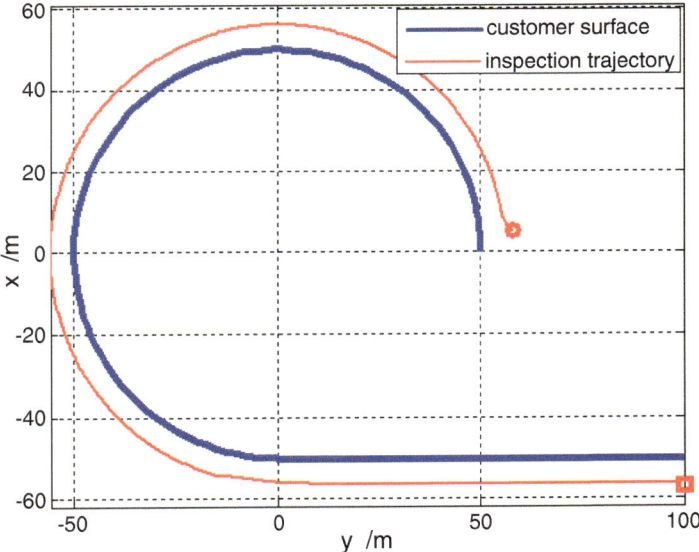

Fig. 4.28 The inspection trajectory of case-2

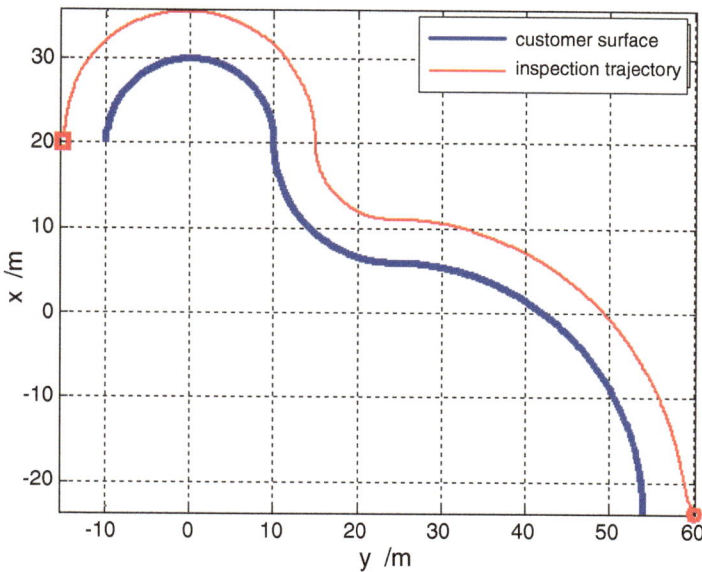

Fig. 4.29 The inspection trajectory with optimized control parameters

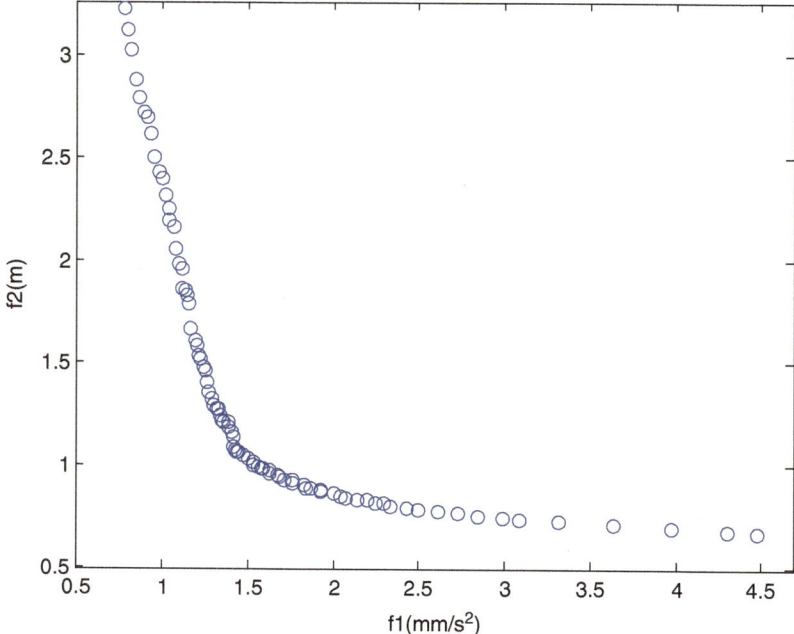

Fig. 4.30 Pareto optimal solution set

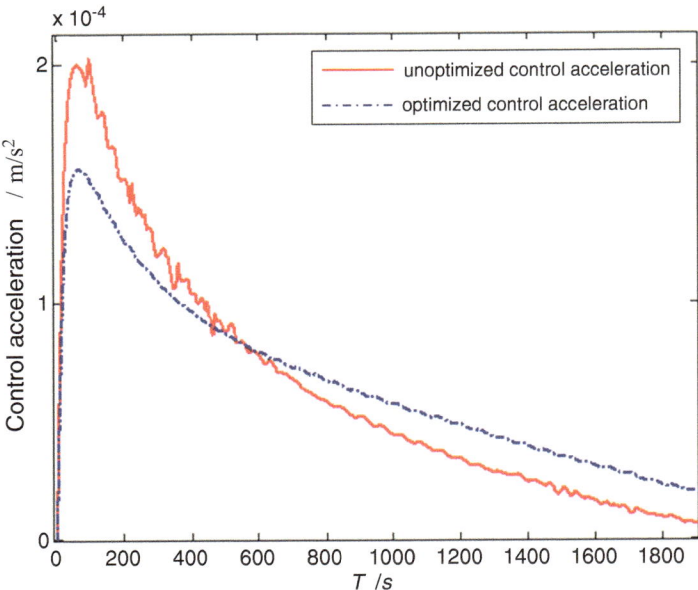

Fig. 4.31 Control acceleration comparison

4.7   IDVD for the Maneuvered Customer

The previous analysis is based on the assumption of a free cooperative customer. However, if the customer is noncooperative, such as a defunct or hostile satellite, its maneuverability must be considered.

4.7.1   Mission Formulation

If the customer is maneuvered, the terminal conditions which are summarized as Eqs. (4.142–4.144) would change.

$$\mathbf{r} = r\mathbf{R}_c^o \mathbf{N}_c^{\text{view}} \tag{4.142}$$

$$\mathbf{R}_c^o \mathbf{N}_c^{\text{view}} \cdot \mathbf{R}_s^o \mathbf{N}_s^{\text{view}} = -1 \tag{4.143}$$

$$\dot{\mathbf{r}} = \boldsymbol{\omega}_c^o \times r\mathbf{R}_c^o \mathbf{N}_c^{\text{view}} \tag{4.144}$$

Equation (4.142) shows the servicer should be along the normal direction of the interested customer surface. Equation (4.143) shows the ideal sensor orientation. Equation (4.144) indicates the customer should stay in the servicer FOV.

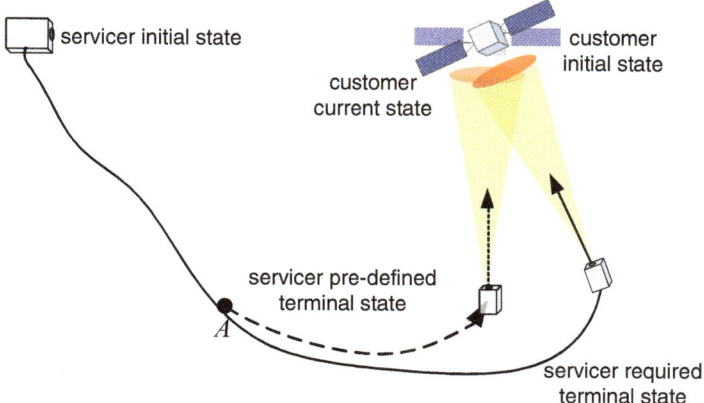

servicer initial state

customer
current state

customer
initial state

servicer pre-defined
terminal state

A

servicer required
terminal state

Fig. 4.32 Influence of attitude maneuver on the servicer

Assume the customer is maneuvered by an impulse thrust $\Delta\mathbf{v}_c$, the servicer relative states change. To satisfy the above terminal constraints, the servicer must maneuver by $-\Delta\mathbf{v}_c$. If the customer attitude also maneuvers, $\mathbf{R}_c^o$, $\mathbf{R}_s^o$, and $\boldsymbol{\omega}_c^o$ would change, as depicted in Fig. 4.32.

To sum up, the customer orbital maneuvers cause the initial conditions to change, while attitude maneuvers cause the terminal conditions to change. Therefore, trajectory planning for a maneuvered customer is essentially a problem of online change of boundary conditions, which requires a fast trajectory generation algorithm.

4.7.2 IDVD Algorithm

IDVD [21], which is now widely used in online trajectory planning of aeronautic vehicles, allows the trajectory to be decoupled in space and time. It takes a reference function of sufficiently higher-order polynomials and matches desired values and potentially derivative values at its terminal to generate a trajectory.

The planning model is derived by selecting certain optimization criteria and taking polynomial coefficients as decision variables. Then, an NLP algorithm is used to generate a solution, with which the reference trajectory and polynomial coefficients can be obtained. Even if the customer states change, partial polynomial coefficients remain the same, and the new approaching trajectory is analytically solved by using current states and expected terminal states as boundary conditions. The advantage of IDVD is that it can make full use of the initial optimization results, and regenerate rapidly a new reference trajectory when boundary conditions change.

The servicer state vector is

$$\mathbf{x} = \left[\mathbf{r}_s^T, \mathbf{v}_s^T, \mathbf{q}_s^T, \boldsymbol{\omega}_s^T \right]_{13 \times 1}^T \tag{4.145}$$

Due to the different format of approximation polynomials for translational motion and rotational motion, the planning models are presented below respectively [22].

4.7.2.1 Translational Motion Planning

The approximation polynomial for the servicer translational trajectory is defined by

$$x(\tau) = a_0 + a_1\tau + a_2\tau^2 + a_3\tau^3 + b_0 \sin\left(\frac{\pi}{2}\tau\right) + b_1 \cos\left(\frac{\pi}{2}\tau\right)$$
$$+ c_0 \sin(\pi\tau) + c_1 \cos(\pi\tau) \tag{4.146}$$

where $\tau \in [0, 1]$, $a_0 \sim a_3$, b_0, b_1, c_0, and c_1 are coefficients.

The first and second derivatives with respect to τ are given as

$$x'(\tau) = a_1 + 2a_2\tau + 3a_3\tau^2 + b_0 \frac{\pi}{2} \cos\left(\frac{\pi}{2}\tau\right)$$
$$- b_1 \frac{\pi}{2} \sin\left(\frac{\pi}{2}\tau\right) + c_0\pi \cos(\pi\tau) - c_1\pi \sin(\pi\tau) \tag{4.147}$$

$$x''(\tau) = 2a_2 + 6a_3\tau - b_0 \frac{\pi^2}{4} \sin\left(\frac{\pi}{2}\tau\right) - b_1 \frac{\pi^2}{4} \cos\left(\frac{\pi}{2}\tau\right)$$
$$- c_0\pi^2 \sin(\pi\tau) - c_1\pi^2 \cos(\pi\tau)$$

Set

$$\lambda(\tau) = \frac{d\tau}{dt} = \lambda_0 + A\tau^2 + (1-\tau)^2 B + \left(1 - (1-\tau)^2\right) C + (1-\tau)^2 D \tag{4.148}$$

where λ_0, A, B, C, and D are coefficients, then

$$t_f = \int_0^1 \frac{1}{\lambda(\tau)} d\tau, \dot{x} = \frac{dx}{d\tau} \frac{d\tau}{dt} = \lambda(\tau) x' \Rightarrow x' = \frac{\dot{x}}{\lambda(\tau)}$$

$$\ddot{x} = \lambda^2(\tau) x'' + \lambda(\tau) \lambda'(\tau) x' \Rightarrow x'' = \frac{\left(\ddot{x} - \lambda(\tau)\lambda'(\tau)\dot{x}\right)}{\lambda^2(\tau)} \tag{4.149}$$

Substituting boundary conditions in t-domain into Eq. (4.149) yields

$$
\begin{aligned}
&x_0 = x|_{\tau=0}, \; x_f = x|_{\tau=1} \\
&x'|_{\tau=0} = \frac{\dot{x}_0}{\lambda(0)}, \; x'|_{\tau=1} = \frac{\dot{x}_f}{\lambda(1)} \\
&x''|_{\tau=0} = \frac{(\ddot{x}_0 - \lambda'(0)\dot{x}_0)}{\lambda^2(0)}, \; x''|_{\tau=1} = \frac{(\ddot{x}_f - \lambda(1)\lambda'(1)\dot{x}_f)}{\lambda^2(1)}
\end{aligned}
\tag{4.150}
$$

Substitute $\tau = 0, 1$ into Eqs. (4.146) and (4.147), combine the resultant equations with Eq. (4.150), and select a_3 and b_1 as decision variables, then

$$
\begin{bmatrix}
1 & 0 & 0 & 0 & 0 & 1 \\
1 & 1 & 1 & 1 & 0 & -1 \\
0 & 1 & 0 & \pi/2 & \pi & 0 \\
0 & 1 & 2 & 0 & -\pi & 0 \\
0 & 0 & 2 & 0 & 0 & -\pi^2 \\
0 & 0 & 2 & -\pi^2/4 & 0 & \pi^2
\end{bmatrix}
\begin{bmatrix}
a_0 \\ a_1 \\ a_2 \\ b_0 \\ c_0 \\ c_1
\end{bmatrix}
=
\begin{bmatrix}
x_0 - b_1 \\
x_1 - a_3 \\
x_0' \\
x_1' - 3a_3 + \pi b_1/2 \\
x_0'' + \pi^2 b_1/2 \\
x_1'' - 6a_3
\end{bmatrix}
\tag{4.151}
$$

The remaining six coefficients can be obtained by solving Eq. (4.151), which generates the translational motion trajectory.

4.7.2.2 Rotational Motion Planning

The fitting polynomial with a quaternion is based on an exponential function

$$
\mathbf{q}(\tau) = \mathbf{q}_0 \prod_{i=1}^{m} \exp\left(\widetilde{\boldsymbol{\omega}}_i \widetilde{\beta}_{i,m}(\tau)\right)
\tag{4.152}
$$

which contains a constant parameter multiplied by a Bezier basis function of a degree m, where [23]

$$
\begin{aligned}
&\widetilde{\beta}_{i,n}(\tau) = \sum_{j=i}^{n} \beta_{j,n}(\tau), \beta_{j,n}(\tau) = \binom{n}{i}(1-\tau)^{n-i}\tau^i \\
&\widetilde{\boldsymbol{\omega}}_i = \ln\left(\widetilde{\mathbf{q}}_{i-1}^{-1}\widetilde{\mathbf{q}}_i\right) \quad i = 1, \ldots, 5
\end{aligned}
\tag{4.153}
$$

The Bezier coefficients are determined by

$$
\begin{cases}
\dfrac{d\mathbf{q}}{d\tau}\bigg|_{\tau=0} = 5\widetilde{\mathbf{q}}_0\widetilde{\boldsymbol{\omega}}_5, \; \dfrac{d\mathbf{q}}{d\tau}\bigg|_{\tau=1} = 5\widetilde{\mathbf{q}}_5\widetilde{\boldsymbol{\omega}}_5 \\[2mm]
\dfrac{d^2\mathbf{q}}{d\tau^2}\bigg|_{\tau=0} = -20\widetilde{\mathbf{q}}_0\widetilde{\boldsymbol{\omega}}_1 + 25\widetilde{\mathbf{q}}_0\widetilde{\boldsymbol{\omega}}_1^2 + 20\widetilde{\mathbf{q}}_0\widetilde{\boldsymbol{\omega}}_2 \\[2mm]
\dfrac{d^2\mathbf{q}}{d\tau^2}\bigg|_{\tau=1} = -20\widetilde{\mathbf{q}}_4\widetilde{\boldsymbol{\omega}}_4\widetilde{\mathbf{q}}_4^{-1}\widetilde{\mathbf{q}}_5 + 20\widetilde{\mathbf{q}}_5\widetilde{\boldsymbol{\omega}}_5 + 25\widetilde{\mathbf{q}}_5\widetilde{\boldsymbol{\omega}}_5^2
\end{cases}
\tag{4.154}
$$

where

$$\dot{\mathbf{q}}(t_0) = \mathbf{q}(t_0)\boldsymbol{\omega}(t_0), \dot{\mathbf{q}}(t_f) = \mathbf{q}(t_f)\boldsymbol{\omega}(t_f)$$

$$\ddot{\mathbf{q}}(t_0) = \dot{\mathbf{q}}(t_0)2\boldsymbol{\alpha}(t_0) + \dot{\mathbf{q}}(t_0)\boldsymbol{\omega}(t_0), \ddot{\mathbf{q}}(t_f) = \dot{\mathbf{q}}(t_f)2\boldsymbol{\alpha}(t_f) + \dot{\mathbf{q}}(t_f)\boldsymbol{\omega}(t_f)$$

$$\widetilde{\boldsymbol{\omega}}_1 = \frac{1}{10}\boldsymbol{\omega}(t_0), \widetilde{\boldsymbol{\omega}}_2 = \frac{\widetilde{\mathbf{q}}_0^{-1}\ddot{\mathbf{q}}(t_0) + 20\widetilde{\boldsymbol{\omega}}_1 - 25\widetilde{\boldsymbol{\omega}}_1^2}{10}, \widetilde{\boldsymbol{\omega}}_3 = \ln(\widetilde{\mathbf{q}}_2^{-1}\widetilde{\mathbf{q}}_3)$$

$$\widetilde{\boldsymbol{\omega}}_4 = \frac{\widetilde{\mathbf{q}}_4^{-1}(\ddot{\mathbf{q}}(t_f) + 20\widetilde{\mathbf{q}}_5\widetilde{\boldsymbol{\omega}}_1 - 25\mathbf{q}_5\widetilde{\boldsymbol{\omega}}_1^2)\widetilde{\mathbf{q}}_5^{-1}\widetilde{\mathbf{q}}_4}{-20}, \widetilde{\boldsymbol{\omega}}_5 = \frac{1}{10}\boldsymbol{\omega}(t_f)$$

$$\widetilde{\mathbf{q}}_0 = \mathbf{q}(t_0) = \mathbf{q}(\tau)|_{\tau=0}, \widetilde{\mathbf{q}}_1 = \widetilde{\mathbf{q}}_0\exp(\widetilde{\boldsymbol{\omega}}_1), \widetilde{\mathbf{q}}_2 = \widetilde{\mathbf{q}}_1\exp(\widetilde{\boldsymbol{\omega}}_2)$$

$$\widetilde{\mathbf{q}}_3 = \widetilde{\mathbf{q}}_5(\exp(\widetilde{\boldsymbol{\omega}}_4)\exp(\widetilde{\boldsymbol{\omega}}_5))^{-1}, \widetilde{\mathbf{q}}_4 = \widetilde{\mathbf{q}}_5\exp(\widetilde{\boldsymbol{\omega}}_5)^{-1}, \widetilde{\mathbf{q}}_5 = \mathbf{q}(t_f) = \mathbf{q}(\tau)|_{\tau=1}$$

$$(4.155)$$

4.7.3   Planning Model

Given $n = 3$ and $m = 5$, the decision variables are

$$a_3, b_1, \lambda_0^r, A^r, B^r, C^r, D^r$$

$$\varepsilon_0^q, \varepsilon_f^q, \lambda_0^q, A^q, B^q, C^q, D^q$$

where (a_3, b_1) and $(\varepsilon_0^q, \varepsilon_f^q)$ determine the space geometrical features of translational and rotational motion trajectory respectively; (λ_0, A, B, C, D) determines the mapping function between t-domain and virtual domain; the superscripts r and q stand for the translational and rotational motion respectively.

Suppose the translational control of the servicer is enabled by thrusters, and rotational control by momentum exchange wheel, then the cost function is

$$J = \int_{t_0}^{t_f} (u_1^2 + u_2^2 + u_3^2 + u_4^2 + u_5^2 + u_6^2)\mathrm{d}t \qquad (4.156)$$

$$\mathbf{u} = \left[\frac{f_x}{f_{x\,\max}}, \frac{f_y}{f_{y\,\max}}, \frac{f_z}{f_{z\,\max}}, \frac{T_x}{T_{x\,\max}}, \frac{T_y}{T_{y\,\max}}, \frac{T_z}{T_{z\,\max}}\right]^{\mathrm{T}} \qquad (4.157)$$

where $\mathbf{f}$ is calculated by Eq. (4.1), and $\mathbf{T}$ is calculated by Eq. (4.109).

The control constraints are

$$\begin{aligned} \mathbf{f}_{\min} \leq \mathbf{f} \leq \mathbf{f}_{\max} \\ \mathbf{T}_{\min} \leq \mathbf{T} \leq \mathbf{T}_{\max} \end{aligned} \qquad (4.158)$$

The time constraint is

$$\lambda(\tau) > 0 \qquad (4.159)$$

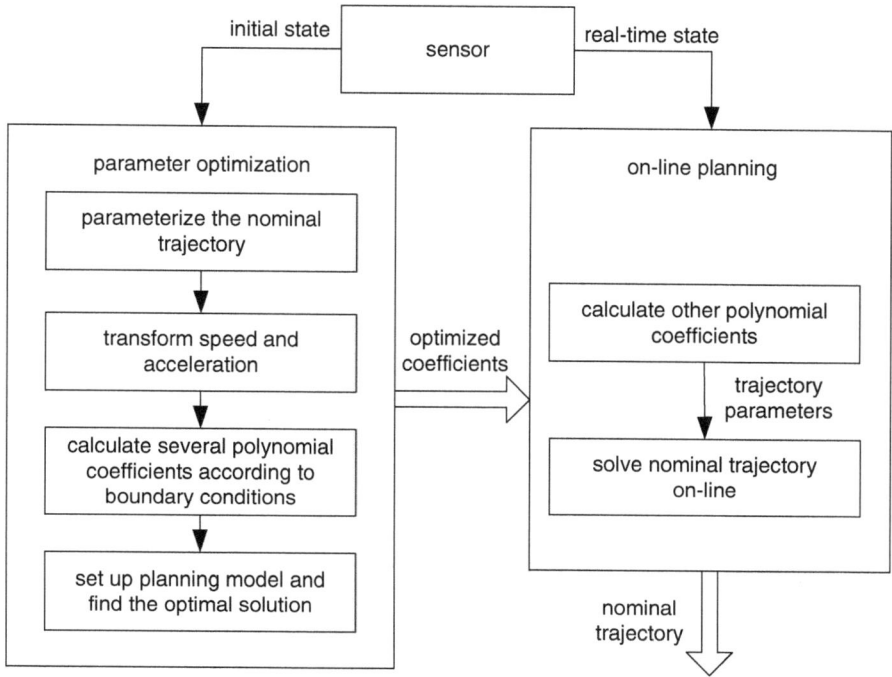

Fig. 4.33 Illustration of solving online trajectory planning

The trajectory constraint is shown as Sect. 4.5.1. The terminal constraints are presented as Eqs. (4.142)–(4.144).

The above planning model is solved by the procedures illustrated in Fig. 4.33.

4.7.4 Numerical Simulation

Assume the customer is a GEO satellite, and the normal vector of interested inspection surface is along the $-x$ axis in the customer body frame. The customer initial attitude quaternion and angular velocity are $[1, 0, 0, 0, 0, 0, 0]^T$. The servicer camera is fixed along its x body axis. The servicer inertial matrix is diag $[2, 2, 2]$kg m^2 and initial states ($t_0 = 0$) are

$$\mathbf{x}_0 = [0, 1000, 100, 0, 0, 0, 0.1423, 0, 0, 0.9898, 0, 0, 0]^T$$

The inspection distance is 500 m and total mission time 4,000 s. The results obtained by IDVD are compared with those by Radau Pseudospectral Method (RPM) (40 collocation points) [24].

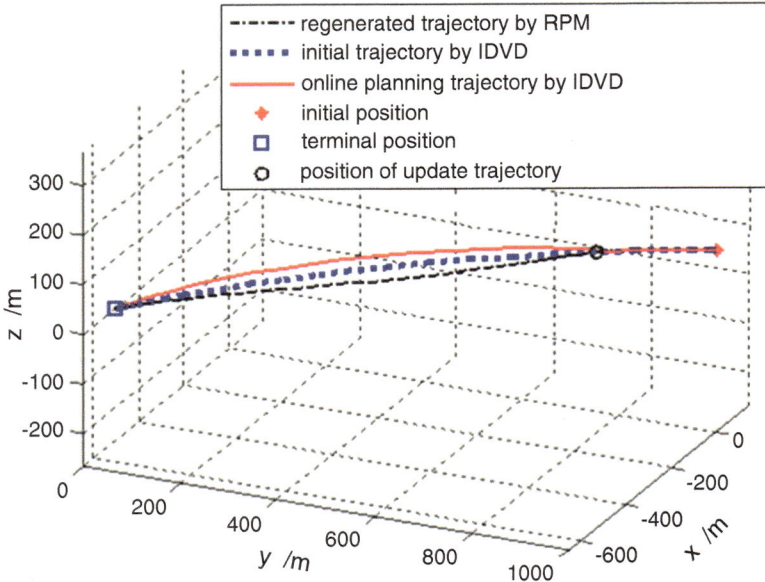

Fig. 4.34 Approaching trajectory of the servicer given a customer orbital maneuver

The optimized decision variables are

$$\lambda_0^r = -1.044\text{e-}3, A^r = 5.546\text{e-}5, B^r = 1.650\text{e-}3$$
$$C^r = 1.489\text{e-}3, D^r = -3.679\text{e-}4, a_3^x = -162.755$$
$$a_3^y = 162.627, a_3^z = 5.082, \lambda_0^q = 2.643\text{e-}5$$
$$A^q = 1.798\text{e-}2, B^q = -1.704\text{e-}2, C^q = -1.757\text{e-}2$$
$$D^q = 1.746\text{e-}2, \boldsymbol{\varepsilon}_0^q = [5.951\text{e-}4, 0, 0], \boldsymbol{\varepsilon}_f^q = \mathbf{0}$$

The reference trajectories under initial boundary constraints are depicted as dotted lines in Figs. 4.34 and 4.35.

Case 1: Assume only the customer orbit changes at $t = 874.7$ s by $\Delta\mathbf{v}_c = [0.2, 0.4, 0.03]^T$ m/s.

The new trajectories obtained by IDVD and RPM are depicted as a solid line and a dash-dotted line in Fig. 4.34. The corresponding cost function is $J = 0.000471$ and the computation time is 2.943 s by RPM, while the cost function is $J = 0.000475$ and the computation time is less than 0.01 s by IDVD.

Case 2: Assume only the customer attitude changes to $[0.9801, 0, 0.1987, 0]^T$, whereas the orbit and total time keep unchanged.

The new trajectory obtained by IDVD is depicted as a thick solid line in Fig. 4.35. The process of attitude maneuver is depicted as in Fig. 4.36. The attitude parameter curves are depicted as in Figs. 4.37 and 4.38.

Since the computation time of servicer trajectory regeneration for each customer attitude maneuver by IDVD is less than 0.01 s, it is viable to deal with the servicer trajectory planning problem considering continuous customer attitude maneuver.

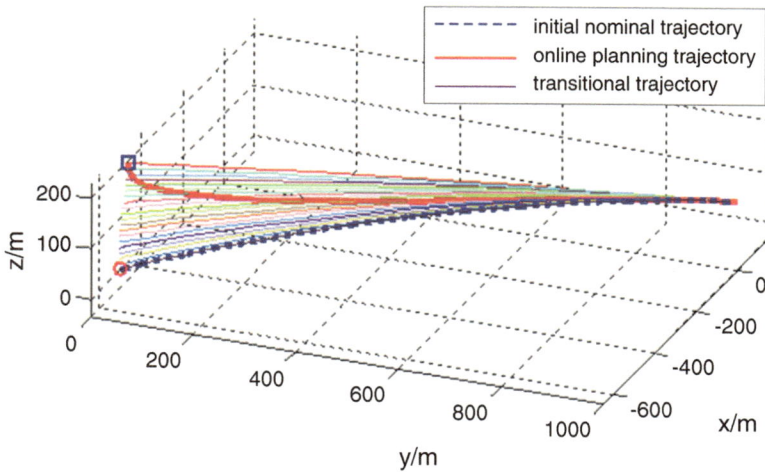

Fig. 4.35 Approaching trajectory of the servicer given a customer attitude maneuver

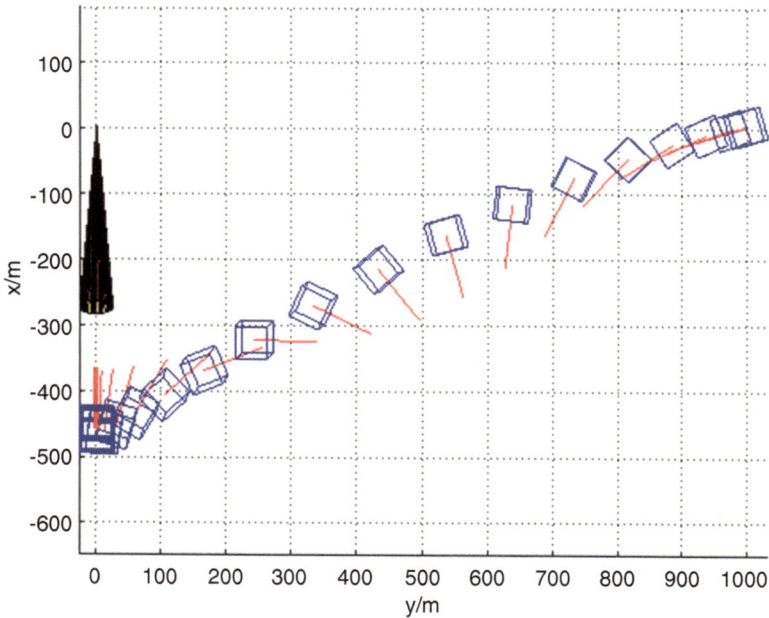

Fig. 4.36 Attitude orientation of the servicer

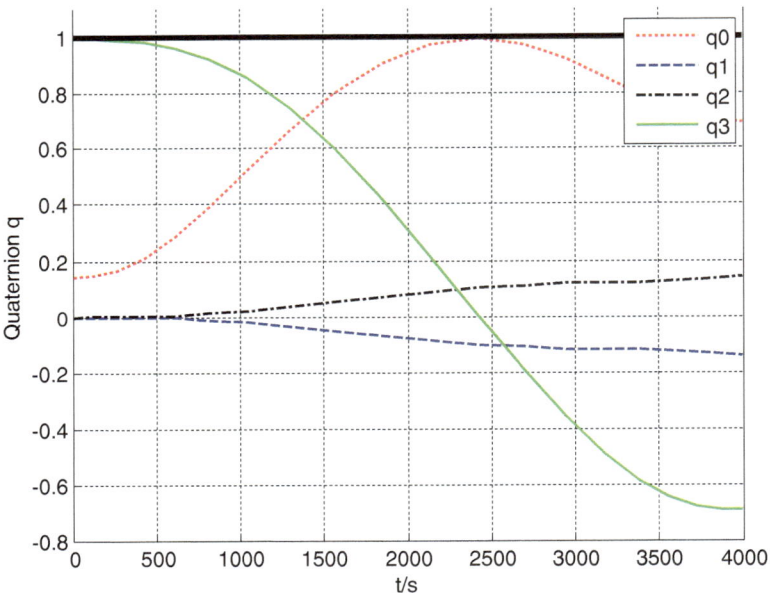

Fig. 4.37 Attitude quaternion curve of the servicer

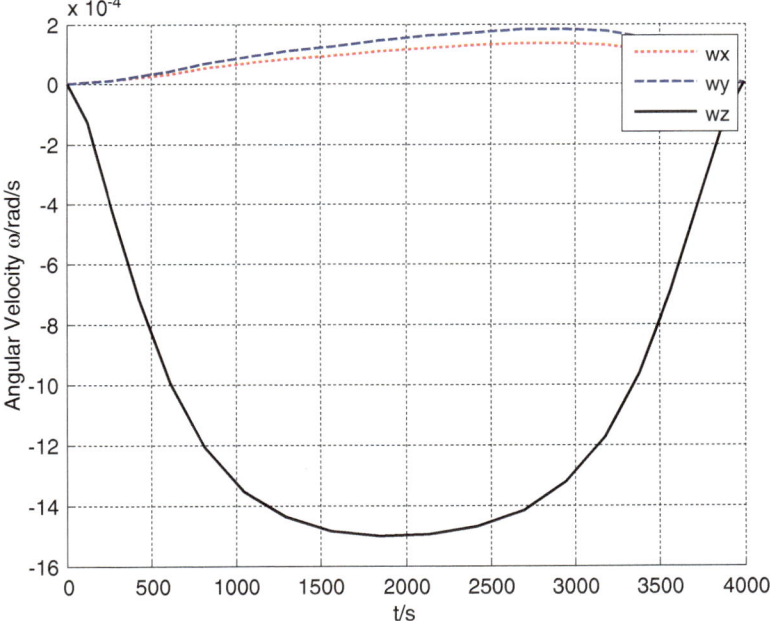

Fig. 4.38 Attitude angular velocity curve of the servicer

References

1. Tang, G. J., Luo, Y. Z., & Zhang, J. (2008). *Mission planning of space rendezvous and docking*. Beijing: Science Press.
2. Straight, S. D. (2002). *Maneuver design for fast satellite circumnavigation*. MS thesis, Air Force Institute of Technology.
3. Yamanaka, K., & Andersen, F. (2002). New state transition matrix for relative motion on an arbitrary elliptical orbit. *Journal of Guidance, Control, and Dynamics, 25*(1), 60–66.
4. Zhu, R. Z. (2007). *Spacecraft rendezvous and docking technology*. Beijing: National Defense Industry Press.
5. Tillerson, M. (2002). *Coordination and control of multiple spacecraft using convex optimization techniques*. MS Thesis, Massachusetts Institute of Technology.
6. Richards, A., Schouwenaars, T., How, J. P., et al. (2002). Spacecraft trajectory planning with avoidance constraints using mixed-integer linear programming. *Journal of Guidance, Control, and Dynamics, 25*(4), 755–764.
7. Schouwenaars, T. (2006). *Safe trajectory planning of autonomous vehicles*. Dissertation, Massachusetts Institute of Technology.
8. Breger, L., & How, J. P. (2007). *Safe trajectory for autonomous rendezvous of spacecraft*. AIAA Guidance, Navigation, and Control Conference and Exhibit, Hilton Head, South Carolina.
9. Zanon, D. J., & Campbell, M. E. (2006). Optimal planner for spacecraft formations in elliptical orbits. *Journal of Guidance, Control, and Dynamics, 29*(1), 161–171.
10. Yu, Q. F., & Shang, Y. (2009). *Videometrics: Principles and researches*. Beijing: Science Press.
11. Kim, S. C. (2006). *Mission design and trajectory analysis for inspection of a host spacecraft by a microsatellite*. MS Thesis, Massachusetts Institute of Technology.
12. Xu, Y., Tatsch, A., & Fitz-Coy, N. G. (2005). *Chattering free sliding model control for a 6 DOF formation flying mission*. AIAA Guidance, Navigation, and Control Conference and Exhibit, San Francisco, California.
13. Darby, C. L., & Rao, A. V. (2009). *A state approximation-based mesh refinement algorithm for solving optimal control problems using pseudospectral methods*. AIAA Guidance, Navigation, and Control Conference, Chicago, Illinois.
14. Rao, A. V., Benson, D., Darby, C. L., Mahon, B., Francolin, C., Patterson, M., et al. (2011). *User's manual for GPOPS version 5.0: A MATLAB software for solving multiple-phase optimal control problems using hp-adaptive pseudospectral methods*. University of Florida.
15. Deslauriersa, A., Englisha, C., Bennettb, C., Ilesa, P., Taylorb, R., & Montpoolb, A. (2006). *3D inspection for the shuttle return to flight*. Orlando, FL: Spaceborne Sensor III.
16. Nellums, R. O., Habbit, R. D., Heying, M. R., Pitts, T. A., & Sandusky, J. V. (2006). *3D scannerless ladar for orbiter inspection*. Orlando, FL: Spaceborne Sensor III.
17. McCamish, S. B. (2007). *Spacecraft guidance strategies for proximity maneuvering and close approach with a tumbling object*. Dissertation, Naval Postgraduate School.
18. McCamish, S. B., Romano, M., & Yun, X. P. (2007). *Autonomous distributed control algorithm for multiple spacecraft in close proximity operations*. AIAA Guidance Navigation and Control Conference and Exhibit, Carolina, Canada.
19. Edwards, C. M. (2008). *Proximity operations of a miniature inspector satellite using emulated computer vision*. MS Thesis, Massachusetts Institute of Technology.
20. Zhang, Q. (2006). *A multi-objective evolutionary algorithm based on decomposition*. Technical Report CSM-450, Department of Computer Science, University of Essex.
21. Yakimenko, O. (2000). Direct method for rapid prototyping of near-optimal aircraft trajectories. *Journal of Guidance, Control, and Dynamics, 23*(5), 865–875.

22. Boyarko, G., Yakimenko, O., & Romano, M. (2010). *Real-time 6DoF guidance for spacecraft proximity maneuvering and close approach with a tumbling object.* AIAA/AAS Astrodynamics Specialist Conference, Toronto, Ontario Canada.
23. Kim, M. J., Kim, M. S., & Shin, S. Y. (1995). *A general construction scheme for unit quaternion curves with simple high order derivatives.* 22nd International Conference on Computer Graphics and Interactive Techniques, Los Angeles, California.
24. Garg, D. (2011). *Advances in global pseudospectral methods for optimal control.* Dissertation, University of Florida.

Chapter 5
Multi-Spacecraft Coordinated Planning

Abstract The development of on-orbit operations technology brings an ever-increasing demand to accomplish a certain mission with multiple smaller and simpler satellites, which requires coordinated planning to achieve and keep a desired configuration. The cyclic pursuit method and the contraction theory are investigated due to their promising features, such as less information exchange and global convergence. First, the cyclic pursuit method is put forward for coordinated trajectory planning, illustrated by cases of constrained circular flyaround, natural elliptical flyaround, flyaround with different orbit, and cubic flyaround. Then, considering on-orbit operations attitude requirements, the contraction theory is applied to 6-DOF synchronized control, illustrated by a four-spacecraft cluster case.

5.1 Problem Formulation

The mission scenario is defined as N coordinated servicers inspecting one customer, which is guaranteed to stay within the servicer configuration by coordinated control.

5.1.1 Dynamic Models

For the 3-DOF case, taking the Hill frame as the reference frame, the relative trajectory dynamic model of servicer i is similar to Eqs. (4.1) and (4.2), see Sect. 4.1.

For the 6-DOF case, the attitude dynamic model should be added together with the relative trajectory dynamic model. To avoid singularity and redundancy,

L. Yang et al., *On-Orbit Operations Optimization: Modeling and Algorithms*,
SpringerBriefs in Optimization, DOI 10.1007/978-1-4939-0838-7_5,
© Leping Yang, Yanwei Zhu, Xianhai Ren, Yuanwen Zhang 2014

Modified Rodrigues Parameters (MRP) $\boldsymbol{\sigma} = [\sigma_1 \quad \sigma_2 \quad \sigma_3]^T$ is utilized to represent the attitude, and then the attitude kinematics model is shown as

$$\dot{\boldsymbol{\sigma}} = \mathbf{Z}(\boldsymbol{\sigma})\boldsymbol{\omega} \tag{5.1}$$

where $\boldsymbol{\omega}$ is attitude angular velocity vector and

$$\begin{aligned}
\mathbf{Z}(\boldsymbol{\sigma}) &= \frac{1}{2}\left(\left(\frac{1 - \boldsymbol{\sigma}^T\boldsymbol{\sigma}}{2} \right) \mathbf{I}_{3\times 3} + \boldsymbol{\sigma}\boldsymbol{\sigma}^T + \tilde{\boldsymbol{\sigma}} \right) \\
&= \frac{1}{4}\begin{bmatrix} 1 + \sigma_1^2 - \sigma_2^2 - \sigma_3^2 & 2(\sigma_1\sigma_2 - \sigma_3) & 2(\sigma_1\sigma_3 + \sigma_2) \\ 2(\sigma_1\sigma_2 + \sigma_3) & 1 - \sigma_1^2 + \sigma_2^2 - \sigma_3^2 & 2(\sigma_2\sigma_3 - \sigma_1) \\ 2(\sigma_1\sigma_3 - \sigma_2) & 2(\sigma_2\sigma_3 + \sigma_1) & 1 - \sigma_1^2 - \sigma_2^2 + \sigma_3^2 \end{bmatrix}
\end{aligned} \tag{5.2}$$

where $\mathbf{I}$ is unit matrix.

And, the attitude dynamic model is similar to Eq. (4.109).

5.1.2 Mission Configurations

Two kinds of mission configurations of multi-spacecraft coordinated operation are exploited in this chapter. One is the line configuration, and the other is the cubic one.

The line configuration usually contains three spacecraft, one customer and two servicers, where the customer is located at the line connecting the two servicers. This configuration could be designed as constrained circular flyaround, natural elliptical flyaround, and flyaround with different orbit, illustrated as Fig. 5.1.

The cubic configuration includes three or more servicers inspecting the customer from different directions. Usually six servicers in a cubic configuration are utilized to fully inspect the customer. If the mission requires more inspection dimensions, the number of servicers can be increased for detailed information, such as the case of eight servicers in Fig. 5.2.

5.1.3 Coordinated Planning

To accomplish the above mission scenario effectively, multi-spacecraft coordinated planning must be addressed for configuration acquisition and maintenance, which has already been sufficiently studied for Distributed Satellite System (DSS).

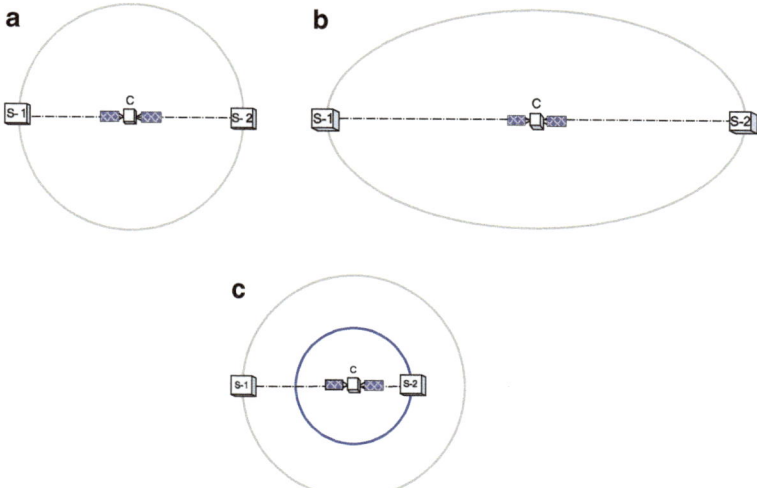

Fig. 5.1 Line configurations for three spacecraft. (**a**) Constrained circular flyaround. (**b**) Natural elliptical flyaround. (**c**) Flyaround with different orbit

Fig. 5.2 Cubic flyaround
formation design

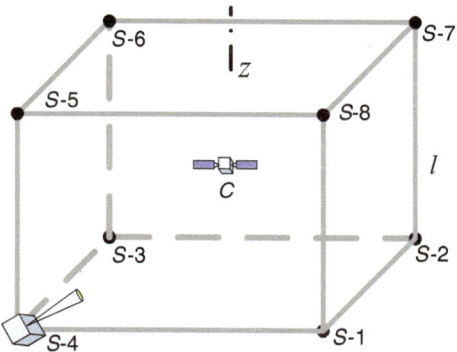

To improve the coordination, several new control methods are investigated, such as the cyclic pursuit method [1–4], which is distributed and requires minimum number of communication links. Therefore, the cyclic pursuit method, which has recently attracted remarkable attention and been verified to be effective and convenient in the context of multi-spacecraft coordinated control, is utilized in Sect. 5.2 to solve the 3-DOF case. In addition, the contraction theory method [5, 6], which could be applied to solving more degrees-of-freedom problem with intrinsic global convergence, is applied to the 6-DOF case in Sect. 5.3.

5.2   Cyclic Pursuit Method

Cyclic pursuit method can address the coordinated control problem in both impulse thrust and continuous thrust mode, while the former corresponds to single-integrator model and the latter to the double-integrator model.

5.2.1   Fundamentals

Consider three-dimensional space n mobile agents (uniquely labeled by an integer $i \in 1, 2, \ldots, n$), where agent i pursues the next $i + 1$, modulo n. Let $\mathbf{x}_i(t) = [x_i(t) \quad y_i(t) \quad z_i(t)]^T \in \mathbb{R}^3$ be the position of agent i.

Case 1: The single-integrator model and corresponding cyclic pursuit control law can be formulated as [3, 4]

$$\dot{\mathbf{x}}_i = k_g \mathbf{u}_i \tag{5.3}$$

$$\mathbf{u}_i^* = \mathbf{R}(\alpha)(\mathbf{x}_{i+1} - \mathbf{x}_i) - k_c \mathbf{x}_i \tag{5.4}$$

where $k_g \in \mathbb{R} > 0$ and $k_c \geq 0$ are determined by the desired configuration; $\alpha \in [-\pi, \pi)$ and $\mathbf{R}(\alpha)$ are the rotation angle and rotation matrix in a datum plane.

Case 2: The double-integrator model and corresponding cyclic pursuit control law can be formulated as

$$\ddot{\mathbf{x}}_i = k_g \mathbf{u}_i \tag{5.5}$$

$$\mathbf{u}_i^* = k_d \mathbf{R}(\alpha)(\mathbf{x}_{i+1} - \mathbf{x}_i) + \mathbf{R}(\alpha)(\dot{\mathbf{x}}_{i+1} - \dot{\mathbf{x}}_i) - k_c k_d \mathbf{x}_i - (k_c + k_d)\dot{\mathbf{x}}_i \tag{5.6}$$

where k_d is the control parameter.

According to the work of Ramirez JL [1, 2], with the above cyclic pursuit controller, the agents can finally converge to a stabilized formation whose center is the initial center of mass of the agents.

1. If $k_c = 0$, the agents starting at any initial condition (except for a set of measure zero) in $\mathbb{R}^{3n}$ exponentially converge:

   (a) If $0 \leq |\alpha| < \pi/n$, to a single limit point.
   (b) If $|\alpha| = \pi/n$, to an evenly spaced circle formation.
   (c) If $\pi/n < |\alpha| < 2\pi/n$, to an evenly spaced logarithmic spiral formation.

2. If $k_c > 0$, the agents converge to formations centered at the origin:

   (a) If $0 \leq |\alpha| \leq \pi/n$, to a single limit point.

(b) If $\pi/n < |\alpha| < 2\pi/n$,

    (i) If $k_c > 2 \sin(\pi/n)\sin(\alpha - \pi/n)$, to a single limit point.
    (ii) If $k_c = 2 \sin(\pi/n)\sin(\alpha - \pi/n)$, to an evenly spaced circle formation.
    (iii) If $k_c < 2 \sin(\pi/n)\sin(\alpha - \pi/n)$, to an evenly spaced logarithmic spiral formation.

5.2.2 Cyclic Pursuit Control Law

According to the dynamic model in Sect. 5.1.1, the multi-spacecraft coordinated control laws based on CyPA, for impulsive thruster and continuous low thruster respectively, are given below.

To characterize the configuration transformation, a nonsingular matrix $\mathbf{T}$ is introduced, then the rotation matrix $\mathbf{R}(\alpha)$ in Eqs. (5.4) and (5.6) can be replaced by $\mathbf{TR}(\alpha)\mathbf{T}^{-1}$.

5.2.2.1 Impulsive Thruster Control Law

The impulsive thruster control law can be derived by Eq. (5.4). Given the nominal velocity of servicer i as $\mathbf{u}_i^*$, and the current velocity as $\mathbf{v}_i$, the control input of servicer i is

$$\mathbf{u}_i = \mathbf{u}_i^* - \mathbf{v}_i \tag{5.7}$$

Substituting Eq. (5.4) into Eq. (5.7) yields

$$\mathbf{u}_i = k_g\left(\mathbf{TR}(\alpha)\mathbf{T}^{-1}(\mathbf{x}_{i+1} - \mathbf{x}_i) - k_c\mathbf{x}_i\right) - \mathbf{v}_i \tag{5.8}$$

where $\mathbf{T}$, α, and k_c determine the desired configuration, and k_g determines the period.

5.2.2.2 Continuous Thruster Control Law

The continuous thruster control law can be derived by Eq. (5.5). Given the current state of servicer i as $\mathbf{x}_i = \begin{bmatrix} x_i & y_i & z_i & \dot{x}_i & \dot{y}_i & \dot{z}_i \end{bmatrix}^T$, if the customer runs in a near-circular orbit, the natural relative acceleration derived by Eq. (4.1) is

$$\mathbf{u}_i^{\text{natural}} = \begin{bmatrix} 2n\dot{y}_i + 3n^2x_i & -2n\dot{x}_i & -n^2z_i \end{bmatrix}^T \tag{5.9}$$

If the customer runs in an elliptical orbit, $\mathbf{u}_i^{\text{natural}}$ derived by Eq. (4.2) is

$$
\mathbf{u}_i^{\text{natural}} =
\begin{bmatrix}
\dot{\theta}^2 x_i + \ddot{\theta} y_i + 2\dot{\theta}\dot{y}_i + \dfrac{\mu}{r_c^2} - \dfrac{\mu(r_c + x_i)}{r_s^3} \\[2mm]
-\ddot{\theta} x_i + \dot{\theta}^2 y_i - 2\dot{\theta}\dot{x}_i - \dfrac{\mu y_i}{r_s^3} \\[2mm]
-\dfrac{\mu z_i}{r_s^3}
\end{bmatrix}
\tag{5.10}
$$

Given the nominal acceleration of servicer i as $\mathbf{u}_i^*$, its control input acceleration is

$$
\mathbf{u}_i = \mathbf{u}_i^* - \mathbf{u}_i^{\text{natural}}
\tag{5.11}
$$

Substituting Eq. (5.6) into Eq. (5.11) yields

$$
\begin{aligned}
\mathbf{u}_i = {} & k_g \left(k_d \mathbf{T} \mathbf{R}(\alpha) \mathbf{T}^{-1}(\mathbf{x}_{i+1} - \mathbf{x}_i) + \mathbf{T} \mathbf{R}(\alpha) \mathbf{T}^{-1}(\dot{\mathbf{x}}_{i+1} - \dot{\mathbf{x}}_i) \right. \\
& - k_c k_d \mathbf{x}_i - (k_c + k_d)\dot{\mathbf{x}}_i) - \mathbf{u}_i^{\text{natural}}
\end{aligned}
\tag{5.12}
$$

where $\mathbf{T}$, α, and k_c determine the desired configuration, k_g determines the period, and k_d is control parameter that determines the convergence speed.

In conclusion, the specific control law is closely related to the desired configuration.

5.2.3 Desired Configuration and Corresponding Control Laws

As stated in Sect. 5.1.2, here we investigate four cases, namely constrained circular flyaround, natural elliptical flyaround, flyaround with different orbit, and cubic flyaround.

5.2.3.1 Constrained Circular Flyaround

Suppose the flyaround plane is x–y plane, the stabilized phase difference will be π. The parameters of constrained circular flyaround [5] are period P, orientation matrix $\mathbf{T}$, and circular radius ρ, which would be utilized to tune k_g, $\mathbf{T}$, and k_c.

1. k_g

The relative position of two servicers can be expressed as

$$
\begin{aligned}
\mathbf{x}_1 &= \begin{bmatrix} \rho \sin\left(2\pi t/P\right) & \rho \cos\left(2\pi t/P\right) & 0 \end{bmatrix}^{\text{T}} \\
\mathbf{x}_2 &= \begin{bmatrix} -\rho \sin\left(2\pi t/P\right) & -\rho \cos\left(2\pi t/P\right) & 0 \end{bmatrix}^{\text{T}}
\end{aligned}
\tag{5.13}
$$

Fig. 5.3 Orientation
of flyaround

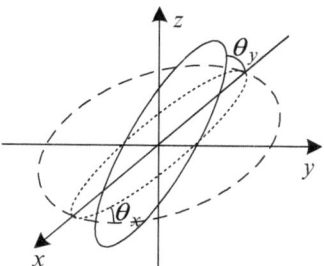

The line speed of constrained circular flyaround satisfies $\left\|\dot{\mathbf{x}}_i^*(t)\right\| = 2\pi\rho/P$;
therefore k_g could be designed based on Eqs. (5.13) and (5.4) as

$$k_g = \frac{\pi}{P \sin \alpha} \tag{5.14}$$

Note that k_g also influences convergence time. To shorten the convergence
time, k_g should be much bigger during the phase of configuration initialization,
and change to a specified value of a given period during the phase of
stabilization.

2. **T**

Taking the x–y plane as datum plane, the desired flyaround plane could be
obtained by rotating x–y plane through θ_x about the x-axis first, and then θ_y
about the y-axis, as shown in Fig. 5.3.
The orientation matrix **T** can be written as

$$\mathbf{T} = \mathbf{T}_y \mathbf{T}_x \tag{5.15}$$

where

$$\mathbf{T}_y = \begin{bmatrix} \cos \theta_y & 0 & -\sin \theta_y \\ 0 & 1 & 0 \\ \sin \theta_y & 0 & \cos \theta_y \end{bmatrix}, \mathbf{T}_x = \begin{bmatrix} 1 & 0 & 0 \\ 0 & \cos \theta_x & \sin \theta_x \\ 0 & -\sin \theta_x & \cos \theta_x \end{bmatrix} \tag{5.16}$$

3. k_c

k_c could be tuned as

$$k_c = -2 \cos \alpha + k_r(\|\mathbf{x}_i - \mathbf{x}_{i+1}\| - 2\rho) \tag{5.17}$$

where $k_r > 0$ is proportional coefficient.

According to the conclusion in Sect. 5.2.1, if $\|\mathbf{x}_i - \mathbf{x}_{i+1}\| > 2\rho$, then $k_c > -2 \cos$
α and ρ decreases; while if $\|\mathbf{x}_i - \mathbf{x}_{i+1}\| < 2\rho$, then $k_c < -2 \cos \alpha$ and ρ
increases. Therefore, the configuration could be stabilized to a ρ-radius circle.

5.2.3.2   Natural Elliptical Flyaround

Based on the control law design in Sect. 5.2.3.1, to save propellant, a natural flyaround ellipse is chosen for the servicers by selecting $\mathbf{T}$ as

$$\mathbf{T} = \begin{bmatrix} 1/2 & 0 & 0 \\ 0 & 1 & 0 \\ z_0 \cos(\phi_z) & z_0 \sin(\phi_z) & 1 \end{bmatrix} \tag{5.18}$$

where ϕ_z is the rotation angle between the current plane and the desired flyaround plane.

According to the conclusion in Sect. 5.2.1, when $t \to \infty$, the relative position of the servicer is

$$\mathbf{x}_i^*(t) = \mathbf{T} \begin{bmatrix} \rho \sin(2\pi t/P + \delta_i) \\ \rho \cos(2\pi t/P + \delta_i) \\ 0 \end{bmatrix} \tag{5.19}$$

where $\delta_i = 2\pi(i-1)/n$ is the evenly distributed phase angle in the initial circular plane, then

$$\mathbf{x}_i^*(t) = \begin{bmatrix} \rho \sin(2\pi t/P + \delta_i)/2 \\ \rho \cos(2\pi t/P + \delta_i) \\ z_0 \rho \sin(2\pi t/P + \delta_i + \phi_z) \end{bmatrix} \tag{5.20}$$

5.2.3.3   Flyaround with Different Orbit

For the above two cases, all the servicers converge to the same flyaround orbit. However, some certain missions require the servicers fly around with different orbit. By adding virtual servicers, the above control law is still valid.

Take two servicers for example, named S-1 and S-2, and virtual servicers S-1′ and S-2′ are added, where S-1′ share the same phase of S-1 and the same radius of S-2, and S-2′ share the same phase of S-2 and the same radius of S-1. Set the position of S-1, S-1′, S-2, and S-2′ as $\mathbf{x}_1$, $\mathbf{x}_1'$, $\mathbf{x}_2$, and $\mathbf{x}_2'$ respectively, then $\mathbf{x}_2' = \frac{\|\mathbf{x}_1\|}{\|\mathbf{x}_2\|}\mathbf{x}_2$ and $\mathbf{x}_1' = \frac{\|\mathbf{x}_2\|}{\|\mathbf{x}_1\|}\mathbf{x}_1$. Set the pursuit strategy as S-1 pursuing S-2′, and S-2 pursuing S-1′, as seen in Fig. 5.4, then the cyclic pursuit control laws for the impulsive thruster are

$$\mathbf{u}_1^* = \mathbf{TR}(\alpha)\mathbf{T}^{-1}\left(\mathbf{x}_2' - \mathbf{x}_1\right) - k_{c1}\mathbf{x}_1 \tag{5.21}$$

$$\mathbf{u}_2^* = \mathbf{TR}(\alpha)\mathbf{T}^{-1}\left(\mathbf{x}_1' - \mathbf{x}_2\right) - k_{c2}\mathbf{x}_2 \tag{5.22}$$

Fig. 5.4 Flyaround
with different orbit

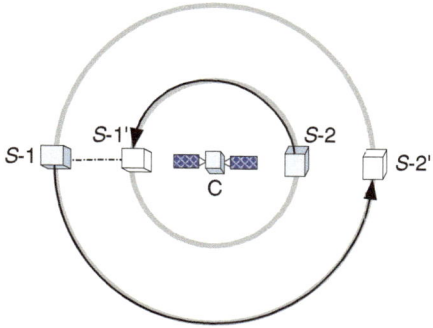

where $\mathbf{T}$ could be similarly tuned as Eq. (5.15) and

$$\begin{cases} k_{c1} = -2\cos\alpha + k_{\rho 1}(\|\mathbf{x}_1\| - \rho_1) \\ k_{c2} = -2\cos\alpha + k_{\rho 2}(\|\mathbf{x}_2\| - \rho_2) \end{cases} \tag{5.23}$$

where $\rho_i (i = 1, 2)$ is the radius of servicer i and $k_{\rho i}$ the corresponding control parameter.

The cyclic pursuit control law for the continuous thruster is

$$\begin{aligned} \mathbf{u}_1^* = {} & k_{\mathrm{d}}\mathbf{TR}(\alpha)\mathbf{T}^{-1}\left(\mathbf{x}_2' - \mathbf{x}_1\right) + \mathbf{TR}(\alpha)\mathbf{T}^{-1}\left(\dot{\mathbf{x}}_2' - \dot{\mathbf{x}}_1\right) \\ & - k_{\mathrm{c}}k_{\mathrm{d}}\mathbf{x}_1 - \left(k_{\mathrm{c}} + k_{\mathrm{d}}/k_{\mathrm{g}}\right)\dot{\mathbf{x}}_1 \end{aligned} \tag{5.24}$$

$$\begin{aligned} \mathbf{u}_2^* = {} & k_{\mathrm{d}}\mathbf{TR}(\alpha)\mathbf{T}^{-1}\left(\mathbf{x}_1' - \mathbf{x}_2\right) + \mathbf{TR}(\alpha)\mathbf{T}^{-1}\left(\dot{\mathbf{x}}_1' - \dot{\mathbf{x}}_2\right) \\ & - k_{\mathrm{c}}k_{\mathrm{d}}\mathbf{x}_2 - \left(k_{\mathrm{c}} + k_{\mathrm{d}}/k_{\mathrm{g}}\right)\dot{\mathbf{x}}_2 \end{aligned} \tag{5.25}$$

$$\dot{\mathbf{x}}_{i+1}' = \frac{\|\mathbf{x}_i\|}{\|\mathbf{x}_{i+1}\|}\dot{\mathbf{x}}_{i+1} \tag{5.26}$$

where k_{g}, k_{c}, and $\mathbf{T}$ could be similarly tuned as in Sect. 5.2.3.1, and k_{d} could be tuned to quicken the convergence of desired flyaround plane.

5.2.3.4 Cubic Configuration

As shown in Fig. 5.2, the eight servicers are classified into two teams. The first team consists of four servicers, written as $s_{\mathrm{m}} = \{\mathbf{x}_1, \mathbf{x}_2, \mathbf{x}_3, \mathbf{x}_4\}$, the second team is $s_{\mathrm{n}} = \{\mathbf{x}_5, \mathbf{x}_6, \mathbf{x}_7, \mathbf{x}_8\}$. In addition, a cooperative team $s_{\mathrm{mn}} = \{\mathbf{x}_3, \mathbf{x}_4, \mathbf{x}_5, \mathbf{x}_6\}$ is defined to cooperative control the first two teams.

The impulse control law of the first team is

$$\mathbf{u}_i^* = k_g \mathbf{R}_m(\alpha)(\mathbf{x}_{i+1} - \mathbf{x}_i) + k_c(\mathbf{x}_i - \mathbf{x}_c) \quad i = 1,2,3,4 \tag{5.27}$$

where $\mathbf{x}_c$ is the center of the configuration consisted of s_m.

The impulse control law of the second team is

$$\mathbf{u}_i^* = k_g \mathbf{R}_n(\alpha)(\mathbf{x}_{i+1} - \mathbf{x}_i) + k_c\left(\mathbf{x}_i - \mathbf{x}_c'\right) \quad i = 5,6,7,8 \tag{5.28}$$

where $\mathbf{x}_c^*$ is the center of the configuration consisted of s_n.

The impulse control law of the cooperative team is

$$\mathbf{u}_{mni}^* = k_g\big(\mathbf{R}_{mn}(\alpha,t)(\mathbf{x}_{i+1} - \mathbf{x}_i) + \mathbf{R}_{mn}^T(\alpha,t)(\mathbf{x}_{i-1} - \mathbf{x}_i)\big) \quad i = 3,4,5,6 \tag{5.29}$$

For the above control laws, the key is to select matrices: $\mathbf{R}_m(\alpha)$, $\mathbf{R}_n(\alpha)$, and $\mathbf{R}_{mn}(\alpha, t)$. Assume the rotating axis of the cube is z-axis, then

$$\mathbf{R}_m(\alpha) = \mathbf{R}_n(\alpha) = \begin{bmatrix} \cos\alpha & \sin\alpha & 0 \\ -\sin\alpha & \cos\alpha & 0 \\ 0 & 0 & 1 \end{bmatrix} \tag{5.30}$$

$$\mathbf{R}_{mn}(\alpha,t) = \begin{bmatrix} \cos(\phi(t)) & 0 & -\sin(\phi(t)) \\ 0 & 1 & 0 \\ \sin(\phi(t)) & 0 & \cos(\phi(t)) \end{bmatrix}$$

$$\times \begin{bmatrix} 1 & 0 & 0 \\ 0 & \cos\alpha & -\sin\alpha \\ 0 & \sin\alpha & \cos\alpha \end{bmatrix} \begin{bmatrix} \cos(\phi(t)) & 0 & -\sin(\phi(t)) \\ 0 & 1 & 0 \\ \sin(\phi(t)) & 0 & \cos(\phi(t)) \end{bmatrix}^T \tag{5.31}$$

where $\dot{\phi}(t) = 2\pi/P$, and P is the flyaround period.

Therefore, the impulse control law of servicer i is

$$\mathbf{u}_i = \mathbf{u}_i^* + \mathbf{u}_{mni}^* - \mathbf{v}_i \tag{5.32}$$

5.2.4   Numerical Simulation

In this section, corresponding numerical simulation cases are implemented to validate the control laws designed in Sect. 5.2.3, where the first three cases are implemented by continuous thruster force with a customer circular orbit of 450 km and the last case by N-times even-interval impulsive thruster force with a GEO customer.

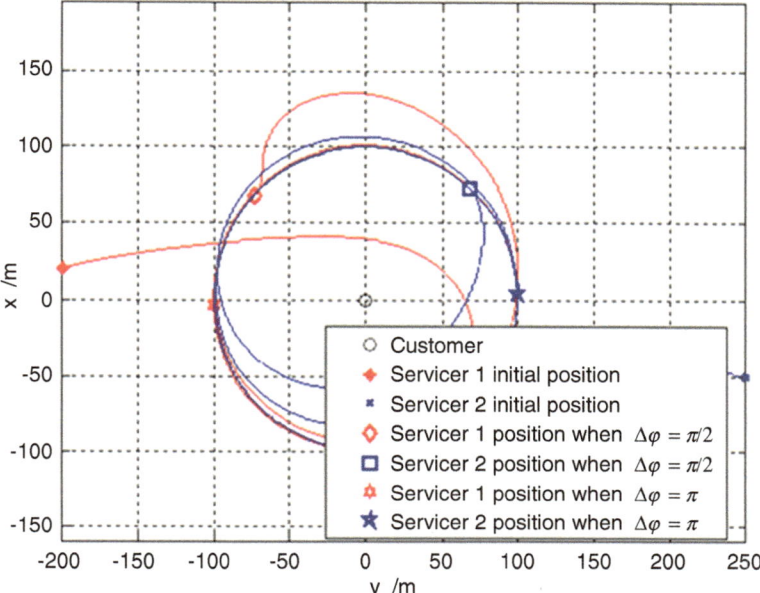

Fig. 5.5 Constrained circular flyaround trajectory

5.2.4.1   Constrained Circular Flyaround

Consider a two-phase flyaround scenario, where the first phase forms a configuration with the phase difference between S-1 and S-2 being $\pi/2$, and the second phase keeps the configuration for two periods and switches to a configuration with the phase difference between S-1 and S-2 being π.

Suppose

$$\mathbf{x}_1 = [20, \, -200, \, -5]^{\mathrm{T}}\mathrm{m}, \dot{\mathbf{x}}_1 = \mathbf{0}$$
$$\mathbf{x}_2 = [-50, 250, 20]^{\mathrm{T}}\mathrm{m}, \dot{\mathbf{x}}_2 = \mathbf{0}$$
$$P = 8,000\,\mathrm{s}, \rho = 100\,\mathrm{m}$$
$$\theta_x = 0, \theta_y = 0$$

Choose the control parameters of phase 1 as

$$\alpha = \pi/3, \quad k_{\mathrm{c}} = 2\sin\left(\pi/4\right)\sin\left(\pi/12\right)$$

and those of phase 2 as

$$\alpha = 5\pi/6, \quad k_{\mathrm{c}} = 2\sin\left(\pi/2\right)\sin\left(\pi/3\right)$$

Then, the flyaround trajectories and corresponding accelerations of two servicers are depicted in Figs. 5.5, 5.6, and 5.7.

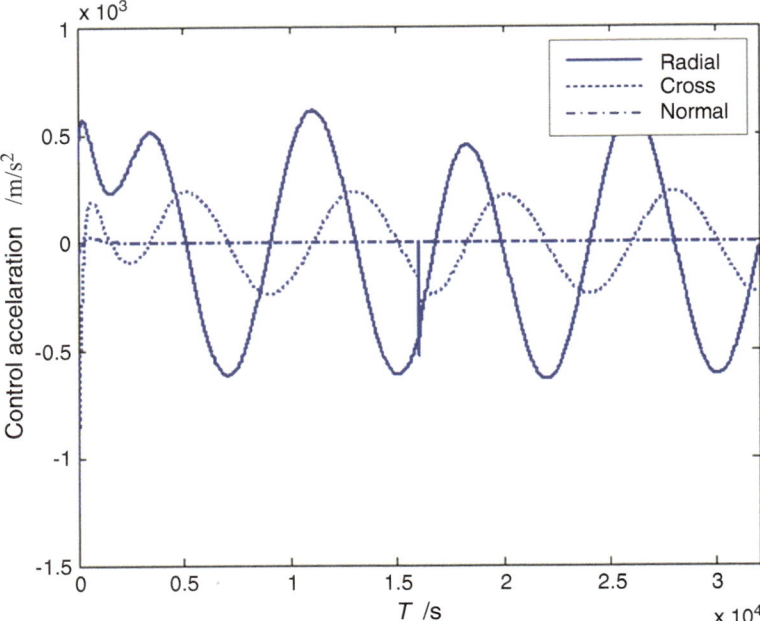

Fig. 5.6 S-1 control acceleration for constrained circular flyaround

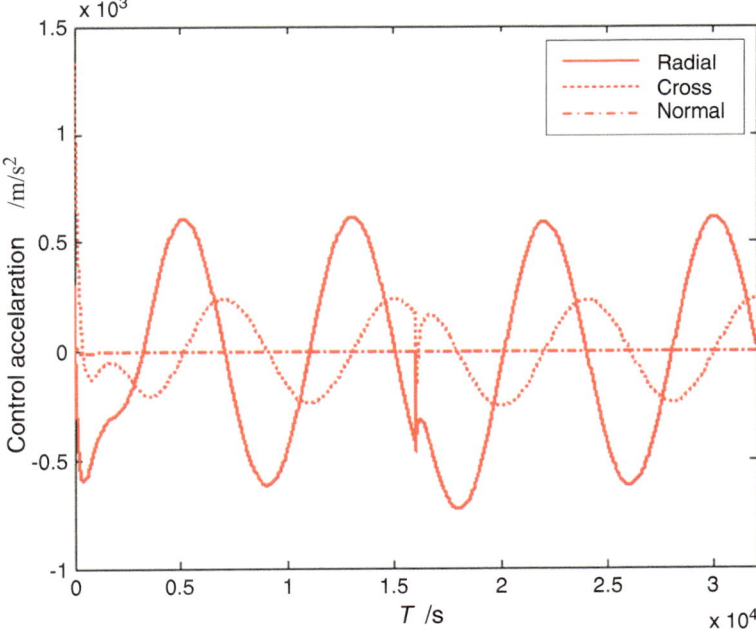

Fig. 5.7 S-2 control acceleration for constrained circular flyaround

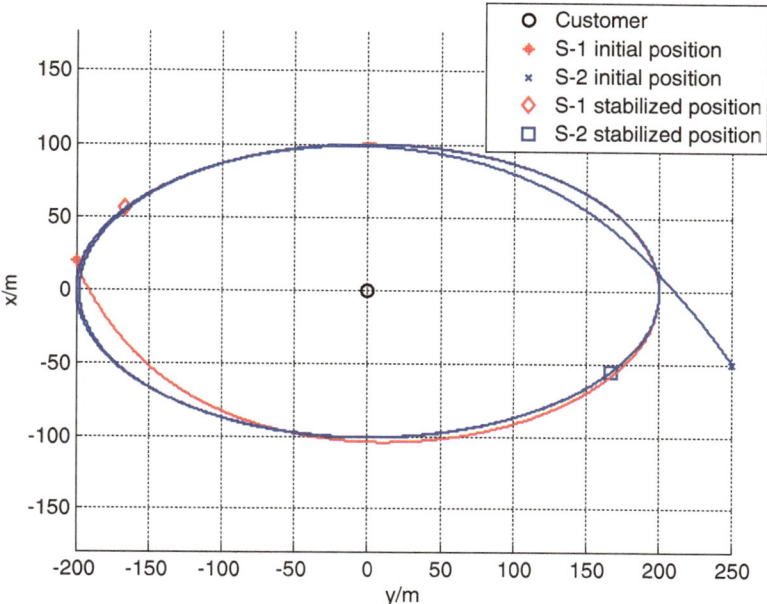

Fig. 5.8 Natural elliptical flyaround trajectory

It can be seen that the two-phase configurations are better produced, and the magnitude of the required control accelerations is less than 10^{-3} m/s². In addition, the convergence time is approximately 4,000 s, which is only one half of the flyaround period, verifying the convergence performance.

5.2.4.2 Natural Elliptical Flyaround

The initial conditions are the same as in Sect. 5.2.4.1. The control parameters are chosen as $\alpha = 5\pi/6$, $k_c = 2\sin(\pi/2)\sin(\pi/3)$, $\mathbf{T} = \text{diag}[1/2, 1, 1]^T$.

Then the flyaround trajectories and corresponding accelerations of two servicers are given in Figs. 5.8, 5.9, and 5.10. Figure 5.8 shows that the two servicers have formed a desired elliptical flyaround configuration. Figures 5.9 and 5.10 show that it takes approximately 4,000 s to form such a configuration, after that the required control accelerations are close to zero, verifying the effectiveness of the designed control law and the characteristics of natural elliptical flyaround.

5.2.4.3 Flyaround with Different Orbit

The initial states are the same as in Sect. 5.2.4.1, and $\theta_x = \pi/3$, $\theta_y = \pi/6$, $\rho_1 = 100$ m, $\rho_2 = 200$ m. Choose the control parameters as $\alpha = 5\pi/6$, $k_c = 2\sin(\pi/2)\sin(\pi/3)$, then the relative trajectories are depicted in Fig. 5.11.

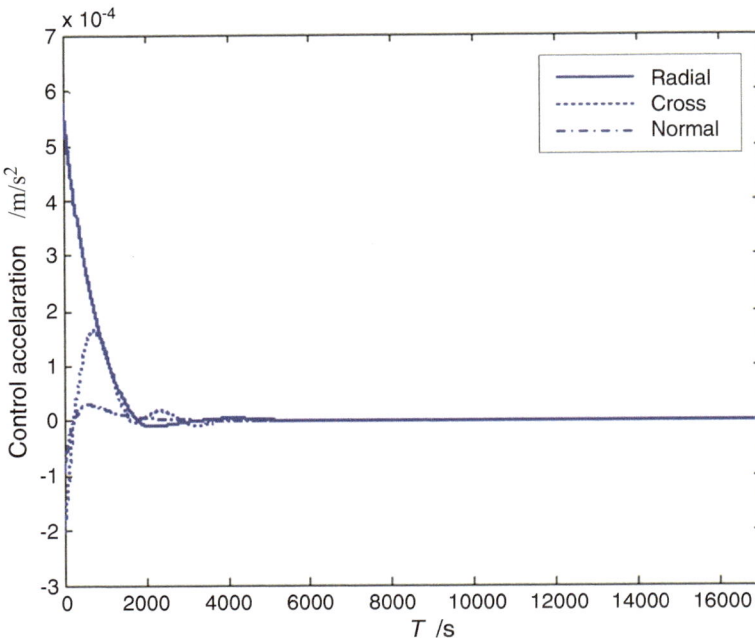

Fig. 5.9 S-1 control acceleration for natural elliptical flyaround

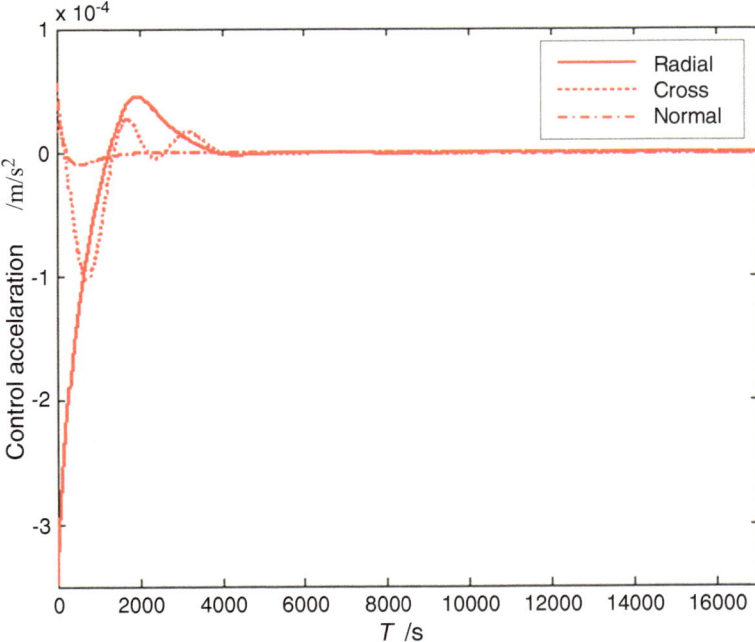

Fig. 5.10 S-2 control acceleration for natural elliptical flyaround

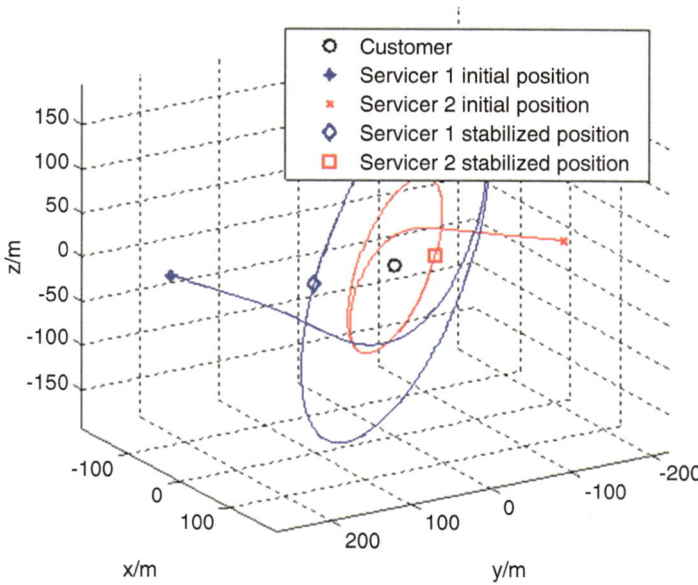

Fig. 5.11 Flyaround trajectory with different orbit

It can be seen that the desired flyaround configuration with different orbits (servicer 2 with a radius of 100 m while servicer 1 with 200 m) is achieved, verifying the effectiveness of the control law.

5.2.4.4 Cubic Flyaround

Assume the customer runs in GEO. The side length of the desired cube is 200 m; the rotating axis is z-axis, and $P = 20, 943$ s. The initial positions of the eight servicers are

$$[-100\,\mathrm{m}, 0, 0]^{\mathrm{T}}, [100\,\mathrm{m}, 0, 0]^{\mathrm{T}}, [0, 100\,\mathrm{m}, 0]^{\mathrm{T}}, [0, -100\,\mathrm{m}, 0]^{\mathrm{T}}$$
$$[0, 0, 100\,\mathrm{m}]^{\mathrm{T}}, [0, 0, -100\,\mathrm{m}]^{\mathrm{T}}, [100\,\mathrm{m}, 100\,\mathrm{m}, 0]^{\mathrm{T}}, [-100\,\mathrm{m}, -100\,\mathrm{m}, 0]^{\mathrm{T}}$$

The initial relative velocities are all zero. Apply an impulse every 10 s. The configurations at $0.1P$ and $0.6P$ are depicted in Figs. 5.12 and 5.13 respectively.

It can be seen that configuration initialization is completed at $0.6P$. The eight servicers converge to a cubic formation, which verifies the effectiveness of the control law.

The velocity increments for configuration initialization and maintenance are given in Table 5.1, which shows the velocity increment of each servicer is acceptable.

Fig. 5.12 Configuration at $0.1P$

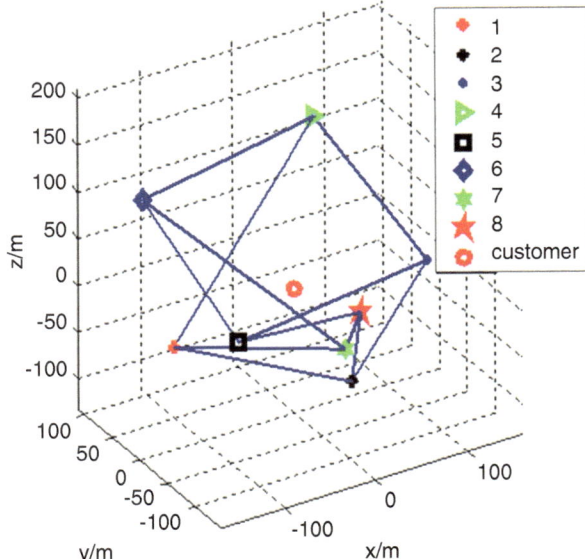

Fig. 5.13 Configuration at $0.6P$

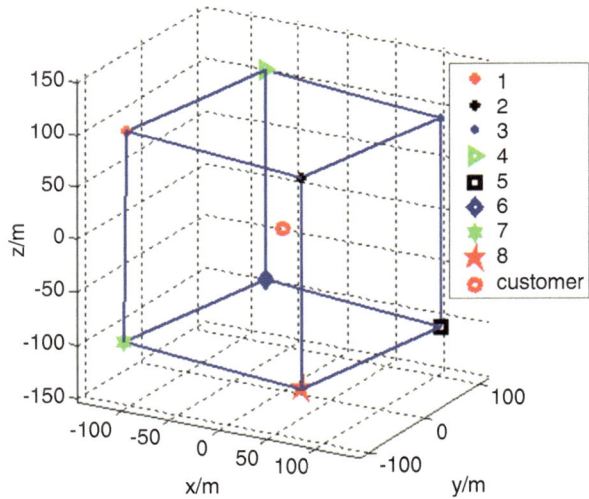

Table 5.1 Simulation results for cubic flyaround

Servicer	Velocity increment for configuration initialization (m/s)	Velocity increment for keeping one period (m/s)	Servicer	Velocity increment for configuration initialization (m/s)	Velocity increment for keeping one period (m/s)
S-1	0.3577	0.6346	S-5	0.3020	0.6345
S-2	0.3469	0.6392	S-6	0.3773	0.6389
S-3	0.3865	0.6345	S-7	0.3216	0.6368
S-4	0.3451	0.6383	S-8	0.3083	0.6403

5.3   Contraction Theory Method

The cyclic pursuit method only addresses the multi-spacecraft trajectory control problem, while on-orbit operations missions usually require 6-DOF control. In this section, we try to use the contraction theory to solve the problem.

5.3.1   Fundamentals

The fundamentals of the contraction theory are summarized as follows [6, 7].
  With system

$$\dot{\mathbf{x}} = \mathbf{f}(\mathbf{x}, t) \tag{5.33}$$

where $\mathbf{f} \subset \mathbb{R}^n$ is a smooth nonlinear function. Define the Jacobian matrix $\partial \mathbf{f}/\partial \mathbf{x}$. If there exists a square invertible matrix $\mathbf{\Theta}(\mathbf{x}, t)$ such that $\mathbf{\Theta}(\mathbf{x}, t)^{\mathrm{T}}\mathbf{\Theta}(\mathbf{x}, t)$ is uniformly positive definite and the matrix

$$\mathbf{F} = \left(\dot{\mathbf{\Theta}}(\mathbf{x}, t) + \mathbf{\Theta}(\mathbf{x}, t)\frac{\partial \mathbf{f}}{\partial t}\right)\mathbf{\Theta}(\mathbf{x}, t)^{-1} \tag{5.34}$$

is uniformly negative definite, then all the system trajectories converge exponentially to a single trajectory.

5.3.2   Unified Model Frame of Dynamic Models

To conveniently apply the contraction theory to the 6-DOF control problem, a unified model frame is introduced to express these dynamic models.
  The state space model of trajectory dynamics can be written as

$$\mathbf{M}_{\mathrm{T}}\ddot{\mathbf{x}}_{\mathrm{T}} + \mathbf{C}_{\mathrm{T}}\dot{\mathbf{x}}_{\mathrm{T}} + \mathbf{Q}_{\mathrm{T}}\mathbf{x}_{\mathrm{T}} = \mathbf{f}_{\mathrm{T}} \tag{5.35}$$

where $\mathbf{x}_{\mathrm{T}} = [x \quad y \quad z]^{\mathrm{T}}$. The matrices $\mathbf{M}_{\mathrm{T}}, \mathbf{C}_{\mathrm{T}}, \mathbf{Q}_{\mathrm{T}}$ are formulated as

$$\mathbf{M}_{\mathrm{T}} = \begin{bmatrix} 1 & & \\ & 1 & \\ & & 1 \end{bmatrix}, \mathbf{C}_{\mathrm{T}} = \begin{bmatrix} 0 & -2n & 0 \\ 2n & 0 & 0 \\ 0 & 0 & 0 \end{bmatrix}, \mathbf{Q}_{\mathrm{T}} = \begin{bmatrix} -3n^2 & 0 & 0 \\ 0 & 0 & 0 \\ 0 & 0 & n^2 \end{bmatrix} \tag{5.36}$$

Similarly, the state space model of attitude dynamics is

$$\mathbf{M}_A(\boldsymbol{\sigma})\ddot{\boldsymbol{\sigma}} + \mathbf{C}_A(\boldsymbol{\sigma}, \dot{\boldsymbol{\sigma}})\dot{\boldsymbol{\sigma}} = \boldsymbol{\tau} \tag{5.37}$$

where

$$\begin{cases} \boldsymbol{\tau} = \mathbf{Z}^{-T}\mathbf{T} \\ \mathbf{M}_A(\boldsymbol{\sigma}) = \mathbf{Z}^{-T}\mathbf{I}\mathbf{Z}^{-1} \\ \mathbf{C}_A(\boldsymbol{\sigma}, \dot{\boldsymbol{\sigma}}) = -\mathbf{Z}^{-T}\mathbf{I}\mathbf{Z}^{-1}\dot{\mathbf{Z}}\mathbf{Z}^{-1} - \mathbf{Z}^{-T}(\mathbf{I}\widetilde{\boldsymbol{\omega}})\mathbf{Z}^{-1} \end{cases} \tag{5.38}$$

Therefore, choosing the system state as $\mathbf{x} = [\mathbf{x}_T^T, \boldsymbol{\sigma}^T]^T$ and combining Eqs. (5.35) and (5.37), the unified 6-DOF dynamic model is derived as

$$\mathbf{M}\ddot{\mathbf{x}} + \mathbf{C}\dot{\mathbf{x}} + \mathbf{Q}\mathbf{x} = \mathbf{f}_c \tag{5.39}$$

where $\mathbf{f}_c = \begin{bmatrix} \mathbf{f}_T^T & \boldsymbol{\tau}^T \end{bmatrix}^T$, and $\mathbf{M}, \mathbf{C}, \mathbf{Q}$ are defined as

$$\mathbf{M} = \begin{bmatrix} \mathbf{M}_T & \\ & \mathbf{M}_A(\boldsymbol{\sigma}) \end{bmatrix}, \mathbf{C} = \begin{bmatrix} \mathbf{C}_T & \\ & \mathbf{C}_A(\boldsymbol{\sigma}, \dot{\boldsymbol{\sigma}}) \end{bmatrix}, \mathbf{Q} = \begin{bmatrix} \mathbf{Q}_T & \\ & \mathbf{0} \end{bmatrix} \tag{5.40}$$

5.3.3 *Application of Contraction Theory*

According to the contraction theory, the 6-DOF controller of servicer i is designed as [8]

$$\mathbf{f}_{ci} = \mathbf{M}_i(\boldsymbol{\sigma}_i)\ddot{\mathbf{x}}_{ri} + \mathbf{C}_i(\boldsymbol{\sigma}_i, \dot{\boldsymbol{\sigma}}_i)\dot{\mathbf{x}}_{ri} + \mathbf{Q}_i(\boldsymbol{\sigma}_i)\mathbf{x}_i - \mathbf{K}_1\mathbf{s}_i + \mathbf{K}_2\mathbf{s}_{i-1} + \mathbf{K}_2\mathbf{s}_{i+1} \tag{5.41}$$

where $i \in [1, \ldots, n-1]$ and $\mathbf{K}_1, \mathbf{K}_2$ are the two positive definite matrices. And $\dot{\mathbf{x}}_{ri}, \ddot{\mathbf{x}}_{ri}, \mathbf{s}_i$ are defined as

$$\begin{aligned} \dot{\mathbf{x}}_{ri} &= \dot{\mathbf{x}}_d + \boldsymbol{\Lambda}(\mathbf{x}_d - \mathbf{x}_i) \\ \ddot{\mathbf{x}}_{ri} &= \ddot{\mathbf{x}}_d + \boldsymbol{\Lambda}(\dot{\mathbf{x}}_d - \dot{\mathbf{x}}_i) \\ \mathbf{s}_i &= \dot{\mathbf{x}}_i - \dot{\mathbf{x}}_{ri} \end{aligned} \tag{5.42}$$

where $\mathbf{x}_d, \dot{\mathbf{x}}_d, \ddot{\mathbf{x}}_d$ are the user-defined profile and its derivative, and $\boldsymbol{\Lambda}$ is a positive diagonal matrix.

Substituting Eq. (5.41) into Eq. (5.39), based on the contraction theory and by choosing an appropriate square invertible matrix $\boldsymbol{\Theta}(\mathbf{x}, t)$, the global convergence of the relative trajectory and attitude could be theoretically validated [1].

5.3.4 Numerical Simulation

Consider four servicers whose inertia matrices are defined as

$$
\mathbf{I}_1 = \begin{bmatrix} 150 & 0 & -100 \\ 0 & 270 & 0 \\ -100 & 0 & 300 \end{bmatrix}, \mathbf{I}_2 = \begin{bmatrix} 100 & 0 & -50 \\ 0 & 150 & 0 \\ -50 & 0 & 250 \end{bmatrix}
$$
$$
\mathbf{I}_3 = \begin{bmatrix} 20 & 0 & -5 \\ 0 & 50 & 0 \\ -5 & 0 & 65 \end{bmatrix}, \mathbf{I}_4 = \begin{bmatrix} 250 & 0 & -150 \\ 0 & 350 & 0 \\ -150 & 0 & 400 \end{bmatrix}
\tag{5.43}
$$

Suppose the customer runs in a circular orbit of 450 km. The servicers' initial relative velocities are all zero, and relative position is

$$
\mathbf{x}_{T1} = [10, 10, 10]^T \mathrm{m}, \quad \mathbf{x}_{T2} = [15, -15, 10]^T \mathrm{m}
$$
$$
\mathbf{x}_{T3} = [-15, 15, -10]^T \mathrm{m}, \quad \mathbf{x}_{T4} = [-10, -10, -10]^T \mathrm{m}
$$

The servicers' initial relative angular velocities are all 0, and the absolute attitude in its body frame is

$$
\boldsymbol{\sigma}_1 = [0.05 \quad -0.1 \quad 0]^T, \boldsymbol{\sigma}_2 = [0.1 \quad -0.1 \quad 0]^T
$$
$$
\boldsymbol{\sigma}_3 = [0.2 \quad -0.1 \quad 0]^T, \boldsymbol{\sigma}_4 = [0.2 \quad -0.2 \quad 0]^T
$$

And the desired relative position and absolute attitude are

$$
\begin{cases} x = 10 \sin(2\pi(0.01)t) \\ y = 15 \cos(2\pi(0.01)t), \\ z = 0 \end{cases} \quad \begin{cases} \sigma_{d1} = 0.1 \sin(2\pi(0.02)t) \\ \sigma_{d2} = 0.5 \sin(2\pi(0.01)t + \pi/3). \\ \sigma_{d3} = 0 \end{cases}
$$

The control gain matrices are given as

$$
\mathbf{K}_1 = 3\mathbf{I}_{6\times6}, \quad \mathbf{K}_2 = \mathbf{I}_{6\times6}, \quad \boldsymbol{\Lambda} = \mathbf{I}_{6\times6}
\tag{5.44}
$$

The simulation results of relative position and velocity, absolute attitude and its corresponding derivative are depicted as in Figs. 5.14, 5.15, 5.16, and 5.17.

A simple analysis reveals that the four servicers can track the designed time-varying relative position/attitude. In addition, the synchronization process is faster than the tracking convergence process, which can also be simply analyzed because the tracking convergence gain $\mathbf{K}_1 - 2\mathbf{K}_2 = \mathbf{I}_{6\times6}$ is smaller than the synchronization gain $\mathbf{K}_1 + 2\mathbf{K}_2 = 5\mathbf{I}_{6\times6}$.

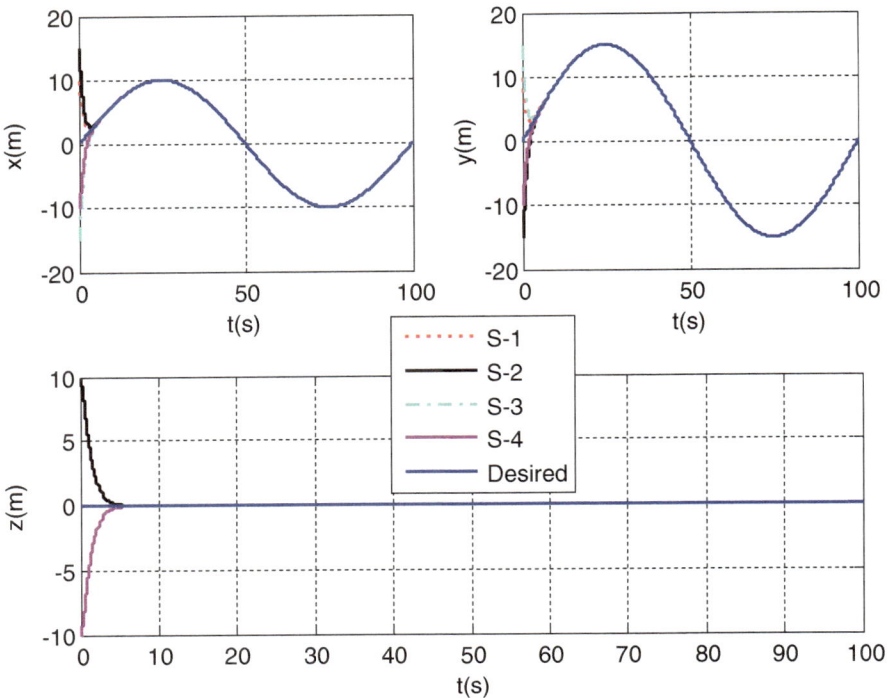

Fig. 5.14 Relative position

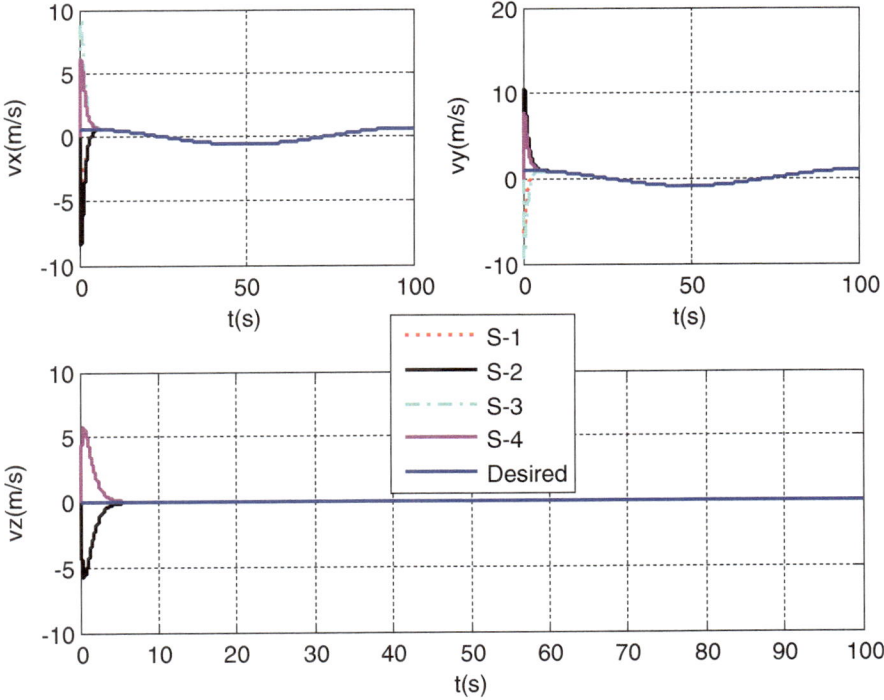

Fig. 5.15 Relative velocity

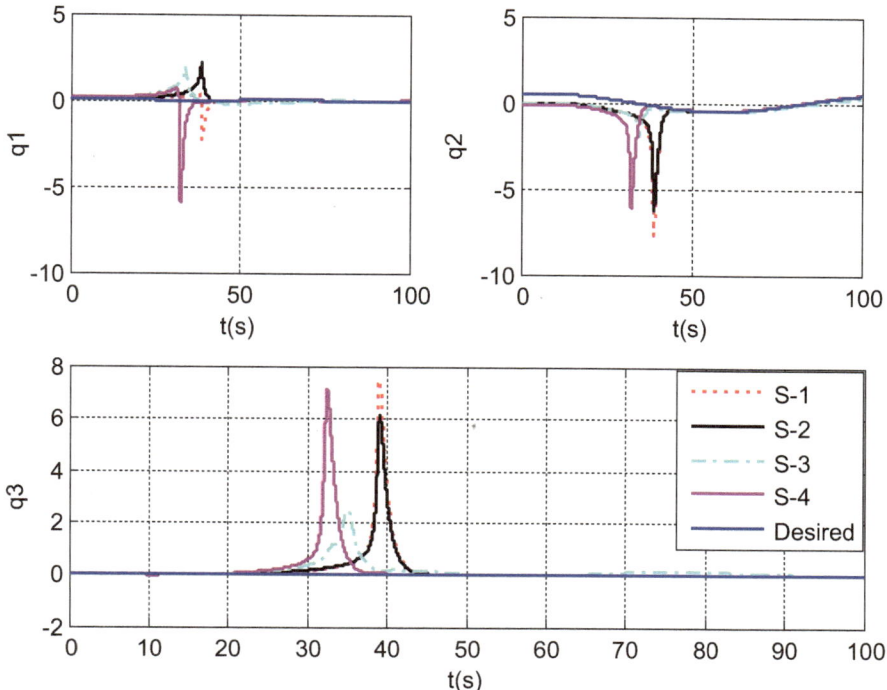

Fig. 5.16 Absolute attitude

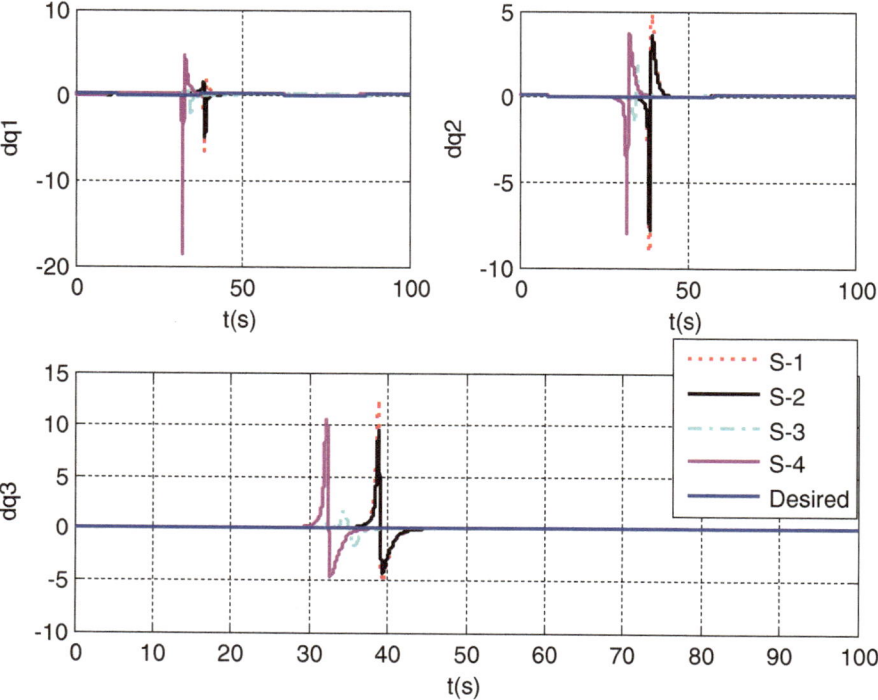

Fig. 5.17 Absolute angular derivative

References

1. Ramirez, J. L., & Frazzoli, E. (2010). *New decentralized algorithms for spacecraft formation control based on a cyclic approach*. Dissertation, Massachusetts Institute of Technology.
2. Ramirez, J. L., Pavone, M., Frazzoli, E., & Miller, D. W. (2010). Distributed control of spacecraft formations via cyclic pursuit: Theory and experiments. *Journal of Guidance Control and Dynamics, 33*(5), 1655–1669.
3. Yang, T., Radice, G., & Zhang, W. H. (2009). *Application of pursuit algorithms for space missions*. IEEE Aerospace Conference, Montana, Bigsky.
4. Yang, T., Radice, G., & Zhang, W. H. (2010). Cooperative control for spacecraft formation reconfiguration via cyclic pursuit strategy. *Advances in the Astronautical Sciences, 134*, 1653–1666.
5. Straight, S. D. (2004). *Maneuver design for fast satellite circumnavigation*. MS Thesis, Air Force Institute of Technology.
6. Lohmiller, W., & Slotine, J. E. (1998). On contraction analysis for nonlinear systems. *Automatica, 34*(6), 683–696.
7. Jouffroy, J., & Slotine, J. E. (2004). *Methodological remarks on contraction theory*. The 43rd IEEE Conference on Decision and Control, Atlantis, Bahamas.
8. Chung, S. J., Ahsun, U., Slotine, J. E., Miller, D. W. (2007). *Application of synchronization to cooperative control and formation flight of spacecraft*. AIAA Guidance, Navigation and Control Conference and Exhibit, Hilton Head, South Carolina.

Index

L. Yang et al., *On-Orbit Operations Optimization: Modeling and Algorithms*,
SpringerBriefs in Optimization, DOI 10.1007/978-1-4939-0838-7,
© Leping Yang, Yanwei Zhu, Xianhai Ren, Yuanwen Zhang 2014